S0-DLP-297

NUCLEAR POWER IN CRISIS

NUCLEAR POWER IN CRISIS

Politics and Planning for the Nuclear State

Edited by
ANDREW BLOWERS and DAVID PEPPER

NICHOLS PUBLISHING COMPANY
New York

CROOM HELM
London & Sydney

Croom Helm Ltd, Provident House, Burrell Row,
Beckenham, Kent, BR3 1AT

Croom Helm Australia, 44-50 Waterloo Road,
North Ryde, 2113, New South Wales

British Library Cataloguing in Publication Data

Nuclear power in crisis: politics and
planning for the nuclear state.
1. Nuclear energy — Political aspects —
Great Britain
I. Blowers, Andrew II. Pepper, David
333.79'24'0941 TK9057

ISBN 0-7099-4034-3
ISBN 0-7099-5048-9 Pbk

First published in the United States of America in 1987
by Nichols Publishing Company, Post Office Box 96,
New York, NY 10024

Library of Congress Cataloging-in-Publication Data

Nuclear power in crisis.

Includes index.
1. Nuclear industry — Government policy — Great Britain.
2. Nuclear industry — Government policy — Case studies.
I. Blowers, Andrew. II. Pepper, David.
HD9698.G72N85 1987 333.79'24'0941 86-23710
ISBN 0-89397-267-3

Printed and bound in Great Britain by Mackays of Chatham Ltd, Kent

CONTENTS

Contents

PREFACE

Until the late 1970s, nuclear policy was largely a matter determined and decided by governments mesmerised by the alchemy of nuclear innovation and the prospect of economic prosperity based on technological progress. The various components of the civil nuclear cycle were presented publicly as discrete issues. The political emphasis was on the glamorous so-called front-end of electricity production while the accumulating problems of nuclear waste were largely ignored or buried in the technical reports in the atomic research complexes. But there has emerged in the 1980s a growing anti-nuclear movement linking together the issues of nuclear energy, reprocessing and radioactive waste and exposing the sinister historical and institutional military connections.

The nuclear issue has now moved beyond the customary secretive decision-making into the sphere of open political conflict. This is apparent in the debate over reactor siting and safety policy as well as the 'dirty end' of nuclear operations, waste management and disposal. As governments strive to pursue the nuclear option they confront a vigorous anti-nuclear movement embracing radicals and conservatives, crossing class boundaries and national frontiers. The opposition now also includes some national governments in Western industrial societies such as Denmark and Sweden, an unthinkable component of nuclear opposition a decade ago. A decisive stage has been reached. Conceivably there will be a pause as the pro-nuclear interests recover their ground before the advance continues. Or the retreat has already begun which will end in the abandonment of the nuclear option. And with the Chernobyl accident a new element in the lexicon of nuclear jargon has emerged, symbolising the catastrophe as a watershed. We must now speak of BC and AC - before and after Chernobyl.

This book was conceived when the signs of a sea change in nuclear politics were already evident. In the UK the Government was already embattled with opponents at the Sizewell inquiry, with media attention at the tribulations of Sellafield and with protesters against inland nuclear waste dumps. In the United States, nuclear power was floundering in the wake of economic scepticism, the accident at Three Mile Island and the developing conflicts over the location of nuclear waste disposal sites. While the book was being written, the concern about the risks to health and the environment were growing and spreading across the nuclear nations. As the book was being finished, Chernobyl gave the world an unmistakable glimpse of the nuclear abyss.

This is the first attempt to bring together a detailed and expert analysis of the events of this period within an overall perspective on the political and policy implications. Various chapters consider the contemporary political issues as governments grapple with the future of nuclear energy, the problems of spent fuel management, of waste disposal and the problems of health and risk they engender. The impact on local communities is viewed statistically and in terms of people's reactions to nuclear questions and their behaviour if something goes wrong. The international dimension is emphasised through comparative chapters. There is a focus on a major future problem - that of decommissioning, which is currently little heeded - and the present problem of the secure replacement of existing sources of energy supply. The opening and concluding chapters identify the impact of the nuclear issue on Western democratic systems and speculate about possible changes in policy as governments attempt to find acceptable and legitimate solutions.

Several of the chapters were first presented as papers at the Annual Conference of the Institute of British Geographers at Leeds in 1985. These, and the other chapters in the book, have been fully revised to take into account more recent developments. Chernobyl, which occurred as we were handing over the manuscript, provided a powerful reinforcement to the arguments presented here.

The editors are extremely grateful to all the contributors, who undertook a constant process of redrafting to ensure that the collection reflects the debate up to the time of publication. We express our special thanks to David Lowry of the Open University, who applied his prodigious knowledge of the nuclear industry to critical purpose on various drafts, and to Tim O'Riordan of the University of East Anglia, whose support and comments encouraged us to sustain the effort needed to complete the volume.

We are also indebted to Eira Halliday of the Open University for dealing so efficiently and willingly with a volume of material coming in from all directions. We hope that all these efforts prove worthwhile in contributing to the nuclear debate on whose outcome the future so much depends.

Andrew Blowers

David Pepper

PART 1 :

THE POLITICAL AND PLANNING CONTEXT

Chapter One

THE NUCLEAR STATE - FROM CONSENSUS TO CONFLICT

Andrew Blowers and David Pepper

1. The Context of the Conflict

The moment of transition

In politics there can sometimes be pinpointed a moment when an issue moves from consensus to conflict. This 'moment of transition' releases latent and dispersed political energy and concentrates it, transforming the issue from a period of relative quiescence to one of active and intense debate. Such a moment can be discerned in the case of nuclear power in Britain in the early 1980s. As the conflict developed so its international context became evident. In 1986 the accident at Chernobyl in the Soviet Union established nuclear power as a major international issue. This book examines the conflict in Britain and its relationship to developments elsewhere in the Western world.

During the autumn of 1983 a series of events occurred in Britain which concentrated public attention on the nuclear industry. The campaign against sea dumping waged by the environmental group 'Greenpeace' and the refusal of the National Union of Seamen to permit the annual dump in the North East Atlantic in July forced the government to accede to the international moratorium on sea dumping in September. In late October the Secretary of State for the Environment announced that two sites on the English mainland, Billingham in Cleveland and Elstow in Bedfordshire, had been chosen as the most promising candidates for the disposal of intermediate and low level radioactive wastes. A few days later, on November 1, Yorkshire TV transmitted its programme, 'Windscale, the Nuclear Laundry' on the national network. This suggested a connection between the activities at the Sellafield reprocessing plant and a high incidence of childhood leukaemias nearby. In the middle of November the environmental group 'Greenpeace', attempting to stop up the pipeline through which radioactive wastes poured from Sellafield into the Irish Sea, discovered a large discharge from the plant, which was later confirmed as an accidental leak by the owners, British Nuclear Fuels Ltd (BNFL). These dramatic incidents occurred when the whole future of the nuclear industry was under scrutiny at the Sizewell 'B' planning inquiry - then nearly in its second year - and also against the background of the coal miners' dispute which lead to a national strike a few months later. The case for nuclear power, hitherto buttressed by the tribulations of the coal industry was, in the space of three months, undermined by growing public and media concern about the dangers of nuclear technology.

Up to this point the nuclear industry had secured its expansion with relatively little opposition. The early nuclear power stations had been accepted with minimal public debate; the accident at Windscale in 1957, when radioactivity had been released into the atmosphere, was largely forgotten; and the opposition in the late 1970s to the proposed Thermal Oxide Reprocessing Plant (THORP) at Windscale, though intense at the time, had failed to impede the industry's progress. The industry was generally perceived to be safe and therefore unproblematic. But a developing public concern for the environment in general became linked, in 1983, with specific anxiety about the dangers from industrially produced radioactivity. The near catastrophe at Three Mile Island (Harrisburg, Pennsylvania, USA) in 1979 had a delayed impact in Britain, but it became a major issue at the Sizewell inquiry into the proposal to build a reactor using similar (but not the same) Pressurised Water Reactor technology (the PWR). As the consequences of the coincidental events of the autumn of 1983 unfolded it became clear that the nuclear industry was engaged in a struggle for credibility. A moment of transition had been reached when the industry, hitherto largely secure from public scrutiny, and enjoying the support of a political consensus favouring its development, was thrust into the public arena: open to criticism and a subject of controversy.

Four elements which provide an explanatory context for the ensuing conflict can be perceived. First, is the element of legitimacy. The development of nuclear energy had enjoyed support from the main political parties and had met little resistance from communities which welcomed the prospect of economic benefits from investment. But there was evidence of a shift in public awareness and attitudes which became evident in the impact of environmental pressure groups, hostile media and opinion polls and local resistance to specific nuclear developments. The THORP inquiry, and the proposals to investigate sites for burying high level wastes in such areas as Galloway and central Wales in the early 1980s, had given premonitions of the impending conflict. Government and the nuclear industry could no longer assume unquestioning support for their activities and they became embroiled in a defence of their record of safety and in a struggle over plans for future expansion. By 1983 the nuclear industry was being challenged on every front and its continued legitimacy was in question.

Related to this is the second element: accountability and control. The nuclear industry exemplifies the concentration of power that exists within the decision making processes of the state. Decisions are effected through a closed, cooperative relationship between industry, the scientific community, trades unions and government, and are endorsed by a compliant legislature and client workforce. Control is vested in government bodies (e.g. the Atomic Energy Authority, AEA) and nationalised industries (Central Electricity Generating Board, CEGB; South of Scotland Electricity Board, SSEB; and BNFL). The public interest in safety is the responsibility of government-appointed bodies such as the Nuclear Installations Inspectorate (NII) and the National Radiological Protection Board (NRPB), which are accountable to the responsible minister. There are also government-appointed advisory bodies, such as the Radioactive Waste Management Advisory Committee (RWMAC) and the Advisory Committee on the Safety of Nuclear Installations (ACSNI). The features of devolved responsibility, administrative discretion and scientific expertise, combined with the secrecy surrounding nuclear activities, make public accountability nominal rather than being a practical feature of decision making. Legislation to implement controls over the industry's activities in practice represents what is acceptable to the industry, and is acceded to by Parliamentary representatives who lack the necessary knowledge of or interest in the issues, or access to crucial decisions. The events of 1983 exposed the closed and secretive nature of decision making and fuelled demands for greater openness and debate. The form of limited accounta bility was no longer acceptable.

The problems of legitimacy and accountability are reflected in the third element of the conflict: the changing political environment. Until 1983 attention had been focused on separate, often localised, issues such as the siting of power stations. The balance of power had favoured the nuclear industry, which faced a fragmented opposition. Resistance had been relatively easy to overcome since the national benefits of investment and local gains from employment outweighed remote risks and environmental impacts. But, as doubts about the economic benefits of nuclear power and the awareness of its economic and environmental costs were strengthened, so opposition began to grow. This was particularly so in the case of nuclear waste disposal, where there were no apparent benefits, only costs, to the host communities. Furthermore there was growing recognition of the integrated nature of the nuclear cycle, which broadened the basis of opposition. During 1983 the developing miners' dispute, the Sizewell inquiry, the Sellafield discharge, the leukaemia scare and the nuclear waste issue could be seen not as isolated events but as interdependent aspects of the nuclear issue. At a broader level still, the issue was becoming an integral part of the environmental concern about resources, public health and even survival. All this encouraged an opposition that transcended social and political divisions and began to develop both a local and a national constituency of protest. The power of the nuclear lobby and its client interests remained formidable, but, it seemed, was no longer impregnable.

This shift in the balance of power was not restricted to the UK but was part of the international political dimension; a fourth contextual element in the conflict. In some countries, notably France (see Chapter 3 in this volume) dependence on nuclear energy had reached a point where retreat seemed unlikely. But elsewhere there was disillusion with nuclear energy, as this book frequently points out. The accident at Three Mile Island had undermined confidence in the industry's safety record and in the ability of governments to cope with catastrophe. The dangers of sea dumping had led to international regulations that took the form of prohibition. Potentially more damaging was the disclosure of an international trade in nuclear technology and materials, particularly weapons grade plutonium, for this undeniably expressed the links between civil and military nuclear power. The prospects of a nuclear Armageddon and its horrifying consequences which were revealed in the debate about the 'nuclear winter' gave renewed impetus to the peace movement, which was spreading across Western Europe and focused, in these years, on the deployment of cruise and Pershing missiles. The conflation of the various components of the nuclear issue on an international scale began to appear irresistible.

Developments in these four elements of conflict - legitimacy, accountability, the political environment and the international dimension - explain the transition from consensus to conflict over nuclear power. The relationship between these elements and the conflict can be assessed through an analysis of the various issues which surfaced in the UK at the end of 1983.

Sellafield - the heart of the conflict

Situated on the Cumbrian coast just west of the English Lake District, the Sellafield complex comprises Calder Hall, in 1956 the first nuclear power station to produce electricity; the Windscale reprocessing works for the spent fuel from the first generation Magnox power stations; and the country's only shallow burial facility for low level radioactive waste, at nearby Drigg. By the early 1980s work was due to commence on the THORP project for reprocessing fuel from the second-generation Advanced Gas-cooled Reactors (AGRs) and from foreign and (if developed) British PWRs. Altogether about 6500 workers are employed in BNFL's reprocessing works, with a further 500 at the adjoining Calder Hall plant, and there are around 3000 construction workers on the site. Sellafield is by far the largest nuclear facility in the country and its relatively labour intensive operation is crucial to the economy of West Cumbria. Annual investment is about £300M and export orders (estimated at £2700M) are a significant element in the national economy. Such economic strength has led to powerful support from successive governments and to the allegiance of its dependent workforce (see Chapter 10 of this volume). By 1983 the plant was confronting two related problems - the scale of its radioactive discharges and the plant's potential links with leukaemia.

The claim that Sellafield is the 'dirtiest

plant in the world' - responsible for 1000 times more radioactive effluent than any other European nation (about 90% of the discharges from Northern Europe) and one million times more than reprocessing plants in the United States - was made at the International Water Tribunal in Rotterdam (Guardian, 3 October 1983). It was later revealed that a quarter of a ton of plutonium had been discharged from the plant into the Irish Sea over the years, and that radioactivity from Sellafield could be traced as far afield as North West Scotland, Scandinavia and even Greenland. Fears were confirmed in November, when an attempt by Greenpeace to block the effluent pipeline was challenged by BNFL. The company secured an injunction against the pressure group that led to a £50,000 fine for contempt (reduced to £36,000 on appeal). But Greenpeace's activities had repercussions, since through them a high level of radioactivity in the vicinity of the pipeline was recorded. BNFL subsequently revealed that owing to an operational error radioactive solvent and crud had been flushed out to sea. This had resulted in a level of discharge much higher than usual, but within the authorised limits.

There were three immediate consequences. First was the discovery that radioactivity had been washed inshore, which led to the complete closure of 200 yards of beach for 24 hours and advice to the public not to use the beaches for 20 miles in each direction. This advice was not withdrawn until August in the following year. Second, an immediate investigation into the incident was ordered by the Government. The outcome of this investigation, published in reports by the Nuclear Installations Inspectorate and the Radiochemical Inspectorate in February 1984 (Health and Safety Executive, 1984, Department of the Environment, 1984), was extremely damaging to BNFL. The reports concluded that a combination of inadequate monitoring, poor management procedures, inadequate information and outdated design and engineering, were responsible for the leak. The plant's director commented that: 'Although the incident did not represent a serious hazard to the Sellafield workforce or members of the public it was a serious event'. The Government's response was to accelerate the renovation programme, to tighten up procedures to prevent a recurrence, and to order the plant to achieve far lower discharge levels. The third consequence was that the Director of Public Prosecutions was asked to investigate the incident.

The following August the DPP decided to prosecute BNFL on the grounds that the company had transgressed the requirement to keep discharges as low as reasonably achievable under the Radioactive Substances Act of 1960, and that it had failed to secure adequate control and to keep adequate records of operations and maintenance and of the location and amounts of radioactive materials, under the Nuclear Installations Act of 1965. After a trial lasting 24 days, the company was convicted in July 1985 on four out of six charges and fined a total of £10,000 with £60,000 costs.

The other major question surrounding Sellafield was the imputed link between radioactive discharges and the high local rates of childhood leukaemias suggested in the Yorkshire TV programme in November 1983. The impact on the public of this programme led to immediate but different responses from the industry and government. BNFL tried to pacify the public by statements emphasising the low levels of radiation from the plant. For instance, it claimed that children in the area would need to ingest 20 lbs of dust per year or sit on a muddy estuary for 500 hours before permitted limits of exposure to radioactivity were reached. By contrast, the Government, anxious to alleviate fears, immediately established an inquiry into possible links between radiation from the plant and childhood cancers. The Black Report (HMSO, 1984a), published the following July, revealed that the leukaemia level at Seascale from 1968 to 1978 was ten times the national average, and in the Millom district was four times the average. However, the numbers of leukaemia cases were very small and, 'An observed association between two factors', said the Report, did not 'prove a causal relationship'. The Report concluded that the hypothesis that childhood cancers were the result of proximity to the plant is 'not one which can be categorically dismissed, nor, on the other hand, is it easy to prove'. The Report further argued that the number of additional deaths in Seascale was about 20 per cent of those to be expected from background radiation, and it was able to offer, on this basis, a 'qualified reassurance' to the local population.

The Report failed to quell public alarm. Leukaemia clusters were being noted in several locations, including Winfrith in Dorset, Aldermaston in Berkshire, and Leiston in Suffolk, which were each close to different types of nuclear installations. The Report's methodology was attacked on the grounds that it had failed to include recently diagnosed cancers; it had failed to note the persistence of the Seascale cluster over the period since Sellafield opened; and it had employed inappropriate statistical tests on the evidence. This latter point, made by several critics (and the main theme of Chapter 11 in this volume), employs the premise that if a probability statistic was used (rather than one based on incidence per 1000 people as used in the Black Report) the leukaemia rate at Seascale would have a one in a million probability of occurring by chance if the latest cases were included. The critics stressed the fact that the Ravenglass estuary near Sellafield has 27,000 times the average amount of plutonium in the environment, and that permitted exposures from Sellafield were twenty times those of the United States, Federal Germany and Japan. They argued that the assertion that a link was 'not proven' was a 'scientifically meaningless statement, since a causal relationship of this nature can never be proved' (Guardian, 27 December, 1984). Later on it was revealed that the Black investigation had been misinformed about the level of uranium discharges into the atmosphere, and it was claimed that the right information about this might have altered the statistical conclusions.

Against this background, developments at Sellafield were receiving close media attention. In December 1983 the THORP development was given planning permission by the local planning authority in return for a Government promise to improve the roads serving the plant: an example of planning gain or compensation for the imposition of externalities on the community. Efforts to reassure the public continued with the offer to local residents of body scans by the National Radiological Protection Board and with the announcement of plans to cut radiation levels to one per cent of their peak 1973 levels by 1991. BNFL concluded an agreement with local trade unions for out-of-court compensation settlements where disease contracted through exposure to radiation might be inferred.

Sellafield was also being linked to broader issues, with allegations that its most important purpose was to produce weapons grade plutonium for British and American defence programmes. A report issued in May 1985 suggested that the plutonium output far exceeded that required for civil purposes, although the precise position remained a matter for conjecture since the operations were not subject to international inspection (Edwards, 1985). Aside from its military connections the high costs , commercial viability and safety of the Sellafield works were beginning to cast doubt on the future of reprocessing. A House of Commons Select Committee conclusion that the THORP project should be scrapped unless the financial and employment consequences were too onerous was leaked to The Times at the end of 1985 and confirmed in the published report in March 1986 (HMSO, 1986a). In early 1986 the safety problems at the plant were underlined by three successive incidents - a discharge of half a ton of uranium into the Irish Sea through what was described as an 'operational malfunction'; a leak of plutonium nitrate within the reprocessing plant, which contaminated several workers; and an escape of contaminated water from a broken drainpipe. These events led to the establishment of an investigation by the Health and Safety Inspectorate and a call for the plant's temporary closure by the European Parliament, amid growing speculation about the long-term survival of the plant.

In retrospect this transformation is extraordinary. The closure of local beaches, constant vilification by the media, the prosecution by the Government of one of its own industries, the series of investigations and reports, had brought Sellafield from relative obscurity into the unrelenting glare of publicity. The industry still enjoyed Government support and the endorsement of its workforce. But, the support was no longer unconditional. The Government had to react to public disenchantment with the risks involved in pursuing the reprocessing option. The workforce was increasingly agitated about the hazards within the plant. Sellafield's legitimacy was now in question. The corporate decision making with limited accountability which had ensured the development of Sellafield had, to an extent, been penetrated by more open, pluralistic, conflict. Sellafield's activities had become a political issue whose outcome was uncertain. It was also a focus of international attention; especially from Ireland, which was concerned about the build-up of radioactivity in the Irish Sea. Thus all four elements responsible for the shift from consensus to conflict were present in the case of Sellafield during these years. And reprocessing was the major source of radioactive waste, which was also a subject of conflict at this time. In the case of nuclear waste the industry and government lacked both a strategy and the prospect of political support from a dependent community.

Radioactive waste: the Achilles Heel of the nuclear industry

In 1976 the Sixth Report of the Royal Commission on Environmental Pollution (HMSO, 1976) had recommended that no further expansion of the nuclear industry should occur until the problem of managing long-lived highly radioactive waste (HLW) had been solved. In the early 1980s attempts to secure suitable sites for exploratory investigation for HLW disposal were called off as a result of local opposition and the problem was effectively deferred. The Government therefore turned its attention to the problem of managing the accumulating intermediate and low level wastes (ILW and LLW). By 1983, with sea dumping abandoned for the time being, the only other available option apparently considered was disposal on land of ILW/LLW. Over the next few years the Government was engaged in a struggle to achieve legitimacy for the land burial option and in an attempt to resurrect the possibility of sea dumping (covered in detail in Chapter 6 in this volume).

In October 1983, the Government announced that two sites had been selected as 'most worthy of detailed consideration' for land disposal of radioactive wastes. These were at Billingham, where a disused mine was identified for deep disposal of longer-lived intermediate wastes (ILW), and at Elstow, where a site owned by the CEGB was earmarked for shallow disposal of short-lived ILW and all types of low level wastes (LLW). The proposal provoked immediate and overwhelming opposition from the two communities, who found support among national pressure groups. It was argued that no consultation had been undertaken, that the land disposal option had not been properly evaluated, and that the decision making process would be severely constrained by concentrating on specific sites rather than on the principles of the strategy. The Government, through its agency the Nuclear Industry Radioactive Waste Executive (NIREX), had identified sites before it had legitimated its policy. Furthermore, the criteria and the process for site selection were unclear and they appeared to contradict the choice of sites. As one critic put it, NIREX had thrown a dart at two convenient walls and then painted in the bullseye. The concerted attack bore fruit when, in March 1984, Imperial Chemical Industries - the major employer in Billingham and the owner of the disused mine - withdrew its cooperation with NIREX and announced its opposition to the proposal. The Government's response, in January 1985, was to abandon the Billingham proposal, thus deferring decisions on the disposal of longer-lived ILW. The debate now focused on shallow burial for short-lived ILW and all LLW. At Elstow the opposition had less dramatic success than at Billingham, though the Government and NIREX were challenged at every opportunity. The Government conceded a need to select up to three other sites in addition to Elstow, so that a comparative site evaluation could be undertaken. Permission for exploratory investigation of these sites would be granted by Parliament, thus delaying a full public inquiry until a proposal for development at a specific site was made.

What had appeared in 1983 to be an unequal struggle between two unprepared communities and the combined power of the nuclear industry and government seemed, two years later, a more equal conflict between NIREX and a broad alliance of local and national interests. The issue had brought together people of all parties and supporters and opponents of nuclear energy and the nuclear deterrent, who felt that the proposed waste disposal strategy was premature and unsafe and that alternatives should be explored. This changing balance of power was recognised by NIREX:

> It is not a contest between equals. On one side there is a nationalised industry with responsibility for supplying the country with a vital commodity. It is accountable to government and Parliament and has to ensure that any public information given out is accurate. On the other side are groups which have no responsibility, no public accountability and are able to propagate such misleading and inaccurate information as they think fit (Letter from NIREX to Bedfordshire Times, 14 February 1985).

Yet it was apparent that the Government was increasingly distancing itself from NIREX, aware that any waste disposal strategy which it formulated must command public approval. What had apparently been NIREX's only strategy in 1983 was now to be subjected to evaluation in which all the available options would be considered.

One of these options was sea disposal. Britain had reluctantly abandoned its sea dumping in the North Atlantic to comply with the directive agreed by the international London Dumping Convention. Britain had also accepted in principle the demand to eliminate its pipeline discharges into the Irish Sea, and to adopt the best available technology for controlling sea discharges as required by the Paris Commission in June 1984. In April 1984, the Holliday inquiry, a joint trades union/government investigation of sea dumping, was established, and it reported in the following December (HMSO 1984b). It concluded that the Government must gain public approval for sea dumping by proving that it was the best option. This conclusion led the Government to initiate its own research into the best practicable environmental options for waste disposal and its findings were published in early 1986 (HMSO, 1986b). Essentially they argued that various options (storage, sea disposal, shallow burial) were all practicable and safe for certain grades of ILW and all LLW. On the basis of an evaluation with varying weightings attributed to minimising cost or risk of environmental impact, and comparing the various options, the Report concluded that sea disposal was the best option for specific wastes whatever the weighting. But it considered that shallow burial on land was the best option for over 90 per cent of LLW and for short-lived ILW. The analysis, methodology and conclusions were regarded by environmentalists as scientifically flawed and politically tendentious. It was also pointed out that the continuation of reprocessing in Britain was assumed throughout the Report.

At about the same time the findings of the House of Commons Environment Committee's investigations into radioactive waste were leaked. The Committee, it was said, had found existing waste management policy to be 'primitive in the extreme', accused the industry of a 'defensive secretiveness', questioned the continuation of reprocessing and urged greater accountability through participation in policy making by environmental groups and the public. These criticisms were substantially confirmed in the published report. The Committee emphasised the backwardness of UK research and development, the lack of coherent policy on radioactive waste management and favoured greater experimental research into solutions such as tunnelling under the sea bed from land (HMSO, 1986a).

By 1986 the political conflict over nuclear waste had become clearly delineated. The nuclear industry was anxious to solve the problem of waste disposal so that the future expansion of nuclear power and reprocessing could be secured. The Government, which had initially backed the NIREX site proposals, had subsequently vacillated in the face of opposition and the acknowledged need to develop and justify an overall strategy. With shallow burial identified as the best practicable environmental option the Government was intent on achieving its legitimation. If it succeeded in this it would merely be a matter of selecting an appropriate site. A public inquiry with terms of reference limited to site-specific issues would enable the Government to divide and rule between communities: eventually settling on that site which was politically the most vulnerable. But the political environment was changing. Ranged against both the industry and the Government were environmental groups and local communities from those areas identified or rumoured as having potential sites. The successful Billingham campaign had been spearheaded by BAND (Billingham Against Nuclear Dumping) and the same acronym (Bedfordshire Against Nuclear Dumping) had been adopted by the group opposing the Elstow proposal. Elsewhere local groups were also emerging in areas thought to be on the short-list of sites (e.g. Humberside Against Nuclear Dumping, (HAND); opposed to the possible site at Immingham on the Humber estuary).

After much delay, in February 1986, the Government announced three further sites which could be investigated as potential locations for the shallow disposal of nuclear wastes. These were at South Killingholme in South Humberside, at Fulbeck in Lincolnshire and at Bradwell in Essex. Together with Elstow these communities were now engaged in a struggle to prevent the legitimation of the shallow burial option. Through unity they hoped to avoid the fragmentation of opposition which would inevitably follow if such a policy were to be adopted. So, the waste disposal issue, like that of reprocessing at Sellafield, was at a stage of transition. In the space of three years it had moved from a position of non-decision making to a position of intense conflict, with a protracted battle in prospect whose outcome was uncertain. Like

Sellafield, this issue was part of the conflict over the whole future of the nuclear industry.

Nuclear energy - the uncertain future

Whereas the problem of reprocessing and radioactive waste management emerged as areas of political conflict after 1983, the future of nuclear energy, on which all stages in the civil nuclear cycle ultimately depend, did not provoke such intense public debate. It was a major issue at the Sizewell 'B' inquiry and was clearly at the heart of the miners' dispute but curiously it did not become a central focus of public attention. There are several possible explanations for this. First, nuclear energy gained early legitimation and is now a familiar feature of the industrial landscape. It has therefore achieved public acceptability. Second, there had apparently been few serious problems with British nuclear power stations up to this point and therefore there had been little evident public anxiety in the communities around them. Third, in consequence, nuclear power was perceived by some as more secure and less environmentally damaging than fossil fuel power production. And fourth, the future of the coal industry compared to that of nuclear energy as an issue in the miners' dispute was masked by the greater attention received by the conflict among the miners and the violence and disorder in the coalfields.

However, there were signs that the conditions of consensus about nuclear energy were beginning to founder in the early 1980s, and that the future of the nuclear industry was a developing source of conflict in Britain and some other Western countries. Here, again, the moment of transition can be discerned.

In the UK the Government's energy policy, which envisaged that a growing share of energy production should come from nuclear stations, had not been legitimated by Parliamentary debate but was - as Chapter 4 of this volume points out - a matter of assumption and assertion. The Sizewell inquiry became, by default, the effective forum for the political debate on energy policy. It was not merely an inquiry into future reactor technology, but an examination of the merits of nuclear power itself. The CEGB envisaged nine gigawatts of new generating capacity by the end of the century, of which two-thirds would be needed to replace existing coal-fired and nuclear plant. Although the proportion coming from nuclear energy was unclear, the CEGB intended to build four or five new nuclear stations at between £5 billion and £6 billion. Some estimates even suggested that nine such stations would be needed by the year 2000. It was likely, therefore, that nuclear power's share, of 17 per cent of electricity supply, would increase at the expense of coal. The major impact would occur in the 1990s, when some of the large coal-fired power stations would be due for decommissioning (see Chapter 2 in this volume).

One of the unresolved questions at Sizewell was the relative costs of the rival power sources. Much depended on assumptions about the future price of coal and oil and predictions of exchange rates. But one study demonstrated that - given a low coal price forecast - the low capital/high running costs of coal would be roughly equal to the high capital-/low running costs of nuclear power. However, restructuring of the British coal industry plus the possibility of cheap foreign coal could give coal a cost advantage and, if this and the social costs of job displacement in the coal industry resulting from a nuclear programme had been taken into account, the overall economic advantage of coal would have been shown to be decisive (Gudgin and Fothergill, 1984). Given these economic uncertainties, the case made at Sizewell for nuclear power would seem to rest on a political judgement, based on a desire to reduce dependence on the miners and to invest in a complex technology with its possible economic spin-offs. The fall in the price of oil undermined the economic case for nuclear power still further. Hence the outcome of Sizewell was certain to cast a deep shadow over future energy policy, with fundamental implications for the price of electricity, the jobs of miners and future national investment strategy.

On the question of safety, the nuclear industry also faced a growing political problem. There were various incidents during this period at some of the ageing Magnox power stations (Trawsfynydd, Hinkley Point 'A', Wylfa, Sizewell 'A') which, though contained, impressed on the public the potential dangers. There were, too, revelations of high discharges of radioactive substances (notably tritium) from Hartlepool and Heysham power stations. The near melt-down at Three Mile Island in 1979 (discussed in Chapter 12 of this volume) had repercussions at the Sizewell inquiry, where it was suggested that the authorities were simply unprepared for a major emergency. This point was underlined by dramatic prognostications about the dimensions of a major catastrophe. These predicted that, in the event of a reactor core melt-down with an easterly wind, land in East Anglia might have to be abandoned for five to fifty years, and a fifth of the UK's annual farm produce would have to be destroyed. In the case of a westerly wind the valuable North Sea fisheries would be contaminated, and the entire farm produce of countries like the Netherlands and Denmark could be wiped out. Although the probability of such an occurrence was argued by the CEGB to be as low as once in 33 million years, the possibility, however, remote, that there could be accidents at any of the growing number of power stations was not reassuring.

This problem of nuclear safety introduces an international dimension which, by the 1980s, was placing a constraint on national energy policy. It was highlightd in August 1984 when the *Mont Louis*, a cargo ship en route from France to the USSR and carrying uranium hexafluoride and some partly refined uranium with traces of plutonium, tritium and strontium, collided with a ferry and sank off the Belgian coast. Although there was no contamination the incident drew attention to the international traffic in nuclear products and the possibilities of the release of radioactive substances in international waters. The following year the decision to develop a reprocessing plant at Dounreay on the northern coast of Scotland to service Europe's fast breeder reactors was announced. It provoked conflict between Caithness and the Highland Regional Council (who supported the scheme for the economic development and security that it would bring) and neighbouring areas (Orkney and Shetland and the Western Isles) who, backed by environmental groups, opposed the potential dangers along sea routes and shorelines. The Government, anxious to evade opposition and to ensure the scheme, called in the planning application and narrowed the terms of the prospective public inquiry to land use and environmental implications rather than - as at Sizewell - encouraging a wide ranging debate on the justification in principle for the project. The Government appeared to think that international commercial imperatives would be achieved by employing tactics which would avoid the need to legitimate its policy. The limited inquiry proposed for Dounreay was a portent of the Government's possible approach to the looming problem of how to handle the inquiry for nuclear waste disposal sites.

In the years following 1983 conflict had developed over a range of nuclear issues. Despite the evidence of accidents and hazards and near disasters, and in the face of growing opposition, the nuclear industry still enjoyed the support of government and the acquiescence of a substantial proportion of the population. Suddenly, in the spring of 1986, the nuclear industry's luck ran out. The grim reality of the accident at Chernobyl in the Ukraine projected the future of the nuclear industry as a major issue of international concern.

Chernobyl: the edge of darkness

At 1.23 am on 26 April 1986, a major explosion, with an accompanying release of radioactivity into the atmosphere, occurred at the fourth reactor at Chernobyl some 60 miles north of Kiev. It appears that an operator erroneously pulled out some control rods, and his attempts to get the situation back under control caused part of the reactor to 'go critical'. The top of the core rapidly heated and the zirconium containers which held uranium fuel reacted with the cooling water to produce hydrogen.

As the hydrogen pressure built up, the core's ceiling cracked and air caused the hydrogen to explode and lift off the roof of the reactor building. The graphite in the core of the reactor had caught fire, and the heat melted part of the nuclear fuel. As the core melted, radioactive materials, notably iodine-131 and caesium-137, were propelled into the atmosphere and fanned by winds at 5000 feet. The accident was first acknowledged to the world two days later when abnormally high levels of radiation were recorded in Sweden some 600 miles away. It took almost two weeks to bring the fire under control by plugging the top with sand and lead and pouring concrete underneath to prevent reaction with the water table. The immediate impact killed nine people and 229 suffered severe sickness. The death toll rose to 25 after six weeks, and no fewer than 18,000 were hospitalised for two days after complaining of radiation symptoms. The town of Pripyat, only two miles from the plant, was evacuated within 36 hours but it was six days before evacuation began at Chernobyl ten miles away, with 40,000 people, and from other settlements within 19 miles of the plant. Altogether, 92,000 people were evacuated.

At the time of the accident winds were blowing towards the north east, carrying the radioactive cloud across the sparsely populated area of Byelorussia to the Baltic republics and Scandinavia. To the south, Kiev and the fertile grain lands of the Ukraine avoided the immediate impact. During the following days the cloud drifted west across Europe, giving rise to radiation levels well above background and, when there was heavy rainfall, reaching levels sufficient to cause alarm. In Poland, with borders only 300 miles from Chernobyl, children were provided with iodine to counteract the radioactive iodine which causes thyroid cancer, and milk sales were banned; in Scotland people were advised not to drink rainwater; in several countries there were bans on the sale of milk and leaf vegetables; and the EEC prohibited the import of fresh vegetables from seven East European countries lying within a 1000 km. radius of Chernobyl. Little information was available about precautions taken within the Soviet Union though in Kiev people were warned to wash vegetables, close windows and prevent children playing outside. There was speculation about the long-term effects. Projections were made about the number of likely deaths from cancer within the Soviet Union, varying from 1000 to 10,000 within a decade. Around Chernobyl it would be necessary to remove vast quantities of topsoil and to prevent contamination spreading into the marshes, rivers and reservoirs supplying the agricultural lands of the Ukraine. The economic effects of the possible loss of grain and electricity supply, and the diversion of resources in the Soviet Union, were incalculable. The ramifications of Chernobyl were world-wide, impinging on every aspect of the nuclear conflict.

Chernobyl posed the most serious threat to the industry's legitimacy. The passivity and indifference of the mass of the population, hitherto an implicit source of that legitimacy, was destroyed by the alarm aroused by the accident. Even before Chernobyl public opinion polls registered little enthusiasm for the expansion of nuclear power (in the UK an NOP poll showed only 11% in favour, an *Observer* poll gave 19%) and after it there were substantial majorities against any further development of power stations (70% in Holland, 73% in Italy and 69% in West Germany). Governments were criticised for their lack of preparedness and information. Emergency evacuation plans were shown to be wholly inadequate. In the USA these proposed a 10-mile evacuation zone around reactors, in Sweden 25 to 50 miles, in the UK a mere 1.5 miles. Yet Chernobyl had required the evacuation of a 19-mile zone, which, transposed to the area around, say, Berkeley, Oldbury or Hartlepool power stations, would involve the evacuation of over a million people. One calculation showed that a radiation cloud drifting south from Sizewell to London, 84 miles away, would result in 24,000 cancers and the evacuation of 3.5 million people, with access banned to large areas for up to 17 years. In a nuclear war, nuclear power stations would be targets and could be vapourised, 'releasing enormous quantities of long-lived radionuclides whose damage will be far more widespread than that from the weapons themselves' (Guardian, 9 May, 1986). Despite efforts by the nuclear industry to emphasise the difference in safety standards and reactor design, it became clear that a range of reactors had basic similarities to Chernobyl and lacked secondary containment systems. The public remained unconvinced, and a BBC TV poll showed that over half those sampled did not accept that an accident similar to Chernobyl could not happen in Britain. The statistical probabilities were not believed and, as one correspondent expressed it, 'What matters about a nuclear accident is not the likelihood of its occurrence but the consequences of its happening' (Tom Burke, The Green Alliance, Guardian, 9 May).

In marked contrast to public apprehension was the strident and defensive reaction of the nuclear establishment. Margaret Thatcher asserted in the House of Commons on May 1 that, 'The record of safety and design, operation, maintenance and inspection in this country is second to none'. In a series of revelations, the pathological secrecy of the nuclear industry was revealed and its pretensions to safety exposed. The twenty-year safety reviews of the Magnox stations were incomplete, and monitoring had been reduced as the nuclear inspectorate was cut back. The failure of the Soviet Union to alert the world about Chernobyl was compared with the UK's delay in revealing the accident at Windscale in 1957. Major problems at Bradwell power station in 1969 were disclosed by Tony Benn, who had been Energy Secretary at the time. In the wake of Chernobyl the UK Government agreed to consider greater openness and the publication of incidents at nuclear facilities. As a further gesture to public opinion it was conceded that intermediate wastes should not be buried in shallow disposal sites. Nonetheless, the Government persisted in its defence of a nuclear energy strategy and in its support for reprocessing at Sellafield and Dounreay. Thus, in the weeks after Chernobyl, demands for greater accountability were met with marginal concessions and a reaffirmation of nuclear policy.

There were signs, however, of a far-reaching transformation in the political environment. Chernobyl ensured that the variety of nuclear issues were conflated. The possible abandonment of nuclear energy had become a central issue requiring a political response. In the UK, conflicts over nuclear power, reprocessing and radioactive waste disposal took on a greater political urgency. The Conservative Government, facing an election within two years and avowedly pro-nuclear, had to surmount formidable obstacles. Any specific decision about the location of a nuclear dump would be deferred until after the election but, meanwhile, the opposition to its waste strategy would continue to prove embarrassing. The publication of the Layfield Report on Sizewell would pose an acute dilemma. If the Report opposed the PWR the Government's nuclear policy was in peril; if it supported the PWR, the Government would be in danger of courting electoral unpopularity. The opposition parties were internally divided. Labour's official line after the accident was to publish safety reviews of existing nuclear facilities; to undertake an economic review of the THORP plant; and to oppose any PWR development. But some of the Left were urging the phasing out of existing power stations and the abandonment of reprocessing at Sellafield and Dounreay. The Liberal-SDP Alliance also nurtured both radical and more cautious attitudes. In other Western countries, too, the political consequences of Chernobyl were dramatic. Denmark, with no nuclear power stations, called on Sweden to close the Barsebaeck power station 12 miles from Copenhagen. Sweden's decision to phase out nuclear power was likely to be reinforced by Chernobyl. The West German Government's support for nuclear power was met with increasing resistance, focused on protests at the site for its new reprocessing plant in Bavaria. France, the largest producer of nuclear energy in Western Europe, maintained its aggressively pro-nuclear line. Its disdain for public opinion was epitomised in the cynical phrase of Remy Carle of Electricité de France, 'You don't ask the frogs when you drain the marsh'. National differences were also exposed in Eastern Europe, with Poland pursuing its plans for nuclear reactors on the Baltic in order to be able to maintain its coal exports, while Yugoslavia abandoned plans for a new reactor near Zagreb. In the USA, where no nuclear reactors had been ordered since 1978, there was unlikely to be any enthusiasm for the resumption of nuclear energy development. Within each country, Chernobyl had dictated a reappraisal of nuclear energy policy reflecting national economic, energy, geographical and political circumstances.

Chernobyl had fully established the international political dimension of nuclear power. Sellafield and the Mont Louis accident had already demonstrated the use of water as a transboundary pathway for radioactive materials. Chernobyl provided the horrifying experience of radioactivity transmitted through the atmosphere. The disaster drew attention to the lack of effective international cooperation on the location of nuclear facilities, design standards, emission levels, inspection, monitoring and control. The problem was especially acute in Europe, with 210 of the world's 382 reactors; 143 of them in Western Europe. A substantial number of nuclear power stations are close to national frontiers, yet neighbouring countries are unable to influence the location, design or safety standards of reactors just across their borders. There is no international coordination of emergency procedures. Bodies such as the International Atomic Energy Agency and the EEC have relied primarily on voluntary coordination and permissive safeguards and limits which allow for interpretation and varying national levels of control. It is conceivable that Chernobyl will underline the necessity for stringent and obligatory standards. But, the ability of national nuclear interests and security to resist international supervision cannot be underestimated so long as nuclear energy and defence retain their menacing mutual dependence.

Before Chernobyl the problem of legitimacy was already being faced on various fronts - reprocessing, nuclear waste management and the future of nuclear energy. Opposition had brought these issues into open debate, raising questions of accountability and shifting the balance of power in the conflict. The international dimension was impinging on national decision making in such areas as sea dumping, reprocessing and plutonium exports. With Chernobyl, the transition from consensus to conflict was complete. It was now conceivable to envisage a collapse of confidence in the industry, a detachment of government support and growing support for a non-nuclear future. The unmistakable glimpse of the abyss delivered by Chernobyl suggested an anti-nuclear consensus rising phoenix-like from the ashes of the reactor. But it was also possible that governments and the nuclear industry would retrench, recover confidence and persist with the expansion of nuclear power. Even after Chernobyl the fascination of nuclear power still captivated governments. The power of nuclear interests would not easily be dislodged. As one critic put it:

> The whole thing is a matter of institutional momentum, plus the fact that the nuclear lobby is one of the most powerful in Whitehall, because of the elite mystique associated with nuclear technology and weapons link (Guardian, 8 July, 1985).

2. Illuminating the Conflict: this Book's Contribution

The nuclear state

The context of this book, then, is that of the struggle between the nuclear industry and its opponents. The elements of this struggle have been defined and illustrated above in the account of some of the events and controversies in the UK during the 1980s. As will now be shown, these elements are further illuminated in the later chapters of the book. But they must first be set in the context of the broader and developing concerns about the nature and direction of Western society which have emerged in the general debate about the environment since the 1960s.

The basis for such concerns is an increasing propensity to see environmental conflicts as manifestations of underlying social and economic processes. When such a perspective is adopted, there is a growing recognition, and not merely on the political left, that inherent tendencies in capitalist economies militate against an environmentally secure future. Thus, there is the 'treadmill of production' (Schnaiberg, 1980) which generates ever-increasing demand for consumer goods and for the resources needed to make them (including energy). There is an inherent drive for high technology of a complex and perhaps dangerous nature, which necessitates and legitimates the concentration of wealth and political power (Albury and Schwartz, 1982; Bookchin, 1980). And there is the tendency towards centralisation and bigness of scale, which creates an alienating environment for work and for living (Schumacher 1973, 1980; Porritt 1984). In such an environment, it seems that ordinary people cannot have any meaningful part in decision making.

The nuclear industry is particularly germane to all of these concerns. Perhaps more than any other environmental controversy, the conflict over nuclear power tends to generate fears about democracy versus state control over people and their environment. Such fears have partly fuelled the elements of the conflict described above. The concept of the 'nuclear state' encapsulates them. The 'nuclear state' is really a vision of a future which, at its worst excesses, is some way off reality. But if the industry successfully retrenches during this time of transition, and secures its continued expansion, then perhaps the development of the nuclear state will also be hastened. In such a state, nuclear energy for both 'peace' and war would be a precious and vital commodity. It would have to be guarded against terrorists and 'dissident' elements, and its secrets would be jealously kept. In an atmosphere of secrecy, the ordinary citizen would not know whether the industry was well or ill-managed, or whether it was safe or dangerous. A whole para-military and military apparatus would protect the industry and its practitioners, while the requirements of security and commercial interests would mean that normal democratic processes of decision making would have to be circumvented. Because nuclear power and weapons are such complex subjects, ordinary people could not anyway be considered as having very much to contribute to the debate. The scientific expert, the politician, and the bureaucrat together would be the decision takers. And nuclear security and high technology would also necessitate large scale centralised organisation. The benefits of such technology and such organisation would accrue mainly to large, often multinational, firms and financial interests - ordinary citizens would not share equally in them. In the nuclear state most of the citizens would be alienated, and, as Marxists would point out, the state as an hierarchical power structure would really be the instrument by which the dominant class maintained its economic and political dominance.

Such concern about a potential Orwellian-style nuclear state is by no means the result of purely idle and unfounded speculation. The Chernobyl disaster brought home the reality that the public in both East and West are, and have been, misled in the interests of the nuclear industry. The accident was the catalyst, also, for a stream of adverse comment about past subterfuge by that industry.

News of the accident itself was withheld by the Russians for several days after it happened: initially it was only the fact that the Swedes picked up high radioactivity levels in the atmosphere that informed the West, and the Soviet people, of what had happened. The Soviet Foreign Ministry later denied that the accident had been deliberately concealed: those who had been confronted with it had not realised the magnitude of the event, and local Ukraine officials had not informed Moscow.

Whereas the Soviet system makes little pretence to openness, the West does. In Britain, for example, Government Ministers quickly took advantage of Chernobyl to castigate the Russians for their secrecy, but opinion polls revealed that most of the public did not believe or trust those Ministers themselves, or the nuclear industry in Britain, to be open and frank. Hugo Young (Guardian, 8 May, 1986) thought they had reason:

> the folk memory is filled more with lies than truth: lies about the Windscale fire (in 1957), evasions about numerous subsequent incidents, official documents spelling out a calculated policy of misinformation and subterfuge ... The nuclear history is one of successive governments declining to trust the public.

Such charges were typical, as were the complaints of an MP that the Government had been over reluctant about revealing the details of three consecutive accidents at the Hinkley Point nuclear power station in autumn 1985 and the string of mishaps at Sellafield in early 1986. And on 4 May

1986, the Observer revealed that an explosion had occurred at Dungeness 'A' nuclear power station the previous March. The CEGB had not admitted to the accident until repeatedly questioned by that newspaper, and it was suggested that the Board's spokesman had been evasive and mendacious. There was nothing particularly new about this kind of incident or the attitude of the officials, but it was significant because the revelations occurred less than a week after the Environment Minister had told the House of Commons: 'There is openness and frankness in this country in dealing with the nuclear industry'. This, of course, was a very moot point - one certain effect of Chernobyl was to expose past secrecy and push the debate yet more into the public domain.

In helping us to understand more fully the elements of the current conflict over nuclear power, many of our authors also have something to say about the nuclear industry which relates to this secretive vision of the nuclear state. Their chapters help us to assess how far along the road we may have travelled towards it, or, by contrast, to what extent the opening up of public debate about nuclear power heralds a more pluralistic framework in which the industry and its opponents are on a more equal footing and the relations between industry and public are more open and subject to genuine democratic processes of decision-making.

The front end of the nuclear cycle - the problems for policy making

Part One, particularly, gives us insights into the planning process, showing how accountability for, and control of, the nuclear industry are at issue, and how secretive, uncoordinated and incremental the process has been and can be. Its shortcomings - and strengths - in relation to democracy in Britain, France and the USA, are evident.

First, Bradbeer (Chapter 2) demonstrates how the relationship between nuclear and other sources of electric power (coal, oil, etc) has never been spelled out in Britain, and how there is no overall policy on the appropriate energy sources to meet the growing regional disparity between energy supply and demand. As an extension of this, both Bradbeer and Armstrong (Chapter 4) show that how many and what types of nuclear power stations are intended has never been unambiguously stated. Such a lack of planning is related to the power structure of the nuclear state. This is a hierarchical, undemocratic 'hegemony', to use O'Riordan's term (in Chapter 13). The members of this power structure are bureaucrats, politicians, technocrats and business people: their nexus is a belief in high technology and material consumption as marks of progress, and in nuclear electricity generation as the finest expression of such progress - which also brings them material rewards. Until recently they have never needed, nor wanted, to formulate coherent plans for national critical scrutiny. And their technological optimism rarely made them concerned about detailed uncertainties over specific outcomes of their intentions. This is hardly surprising, given the prevailing climate of near-euphoria during the 1950s and '60s, when the nuclear industry was born. Boyle and Robinson (Chapter 3) describe a state, France, where this powerful nuclear hierarchy has had its way for many years, with little opposition. In Britain and the USA, however, as Fernie and Openshaw point out in Chapter 5, it is no longer so easy. The rise of public concern about the machinations of the nuclear state has led to the political challenge by many dissenting groups which has been discussed above.

The events reveal that, when challenged, the government/nuclear complex has either attempted to define its policy for energy at specific local siting inquiries, like Sizewell (which is a manifestly unsuitable forum for defining and legitimising such policies, or for challenging them), or it has fallen back on the reflex action of not allowing overall policy to be enunciated, discussed or challenged at all - as was the case with the 1986 inquiry into reprocessing at Dounreay, where the policy defined at the Windscale Inquiry of 1977 was assumed still to operate.

The nuclear establishment's reaction to challenge has, therefore, inevitably thrown a spotlight on the decision making process in nuclear matters. Fernie and Openshaw draw attention to the inadequacies of procedures to deal with public consultation. The UK appears to be in a half-way position between the relatively open procedures of the US and the closed process in France. They argue that there should be a move towards more openness which (as Armstrong and O'Riordan suggest) emphasises the need for greater accountability to be <u>seen</u> to operate if the nuclear industry is to succeed in the 'struggle for legitimacy'. But even the apparently democratic forum of the British public inquiry is far from allowing and encouraging full accountability, as most people who have expressed concern over any environmental issue will testify. At the Sizewell Inquiry, the imbalance in the nature of the conflict was dramatised. The nuclear industry spent millions of pounds of public money to make its case; objectors received no support from public funds. Furthermore, issues like safety, nuclear proliferation, and the connection between nuclear power and weapons were all denied an open discussion, for reasons which Armstrong describes. In 1986, the refusal by Friends of the Earth and the Town and Country Planning Association to participate in the Dounreay Inquiry underlined the growing disillusionment of environmental groups which started in 1977 with the Windscale Inquiry (where FoE alleged that the inquiry was 'fixed' in advance) and hardened in 1983 with Sizewell (where FoE was advised not to participate, and did so only despite huge misgivings).

Many see the public inquiry simply as a retrospective legitimation process. Such a view was encouraged by actions like the Central Electricity Generating Board's purchase of £12m-worth of equipment for a PWR in February 1984 - over two years before the Inspector was to pronounce on whether approval was actually to be given for a PWR. The CEGB argued that this action would avoid incurring extra costs later on, and ensure that developments, if approved, would be speedy.

Reading Boyle and Robinson's account of the French nuclear industry - beset by few 'problems' of dissent - there appears to be a negative correlation between the degree of openness of decision making and the coherence and vigour of the nuclear industry. However, lack of proper planning and a too-powerful nuclear establishment have created specific problems. France's industry produces an oversupply of nuclear electricity, and it cannot encourage enough demand - its 'cornucopian' ideology (Cotgrove, 1982) combined with nationalistic overtones which have promoted a pro-growth mentality that must be satisfied have rendered it quite unable to accommodate to the realities of quasi-permanent economic recession in the 1980s.

Part One leaves us, then, with the impression that the components of the nuclear state may exist in embryo. There is, first, a political-technical-commercial complex; second, an ideology of technological optimism and cornucopian thinking; third, a climate of secrecy, unequal access to decision making and the information required to mount credible opposition: and, fourth, there is an underlying tendency from the nuclear establishment to prevaricate and to argue illogically.

The back end of the nuclear cycle: the focus of protest

These characteristics of the nuclear establishment - technological optimism, belief in the primacy of scientific expertise in decision making, a disinclination towards public participation and accountability, and underlying prevarication and tendency to error - make this establishment an excellent example of 'technocentrism' (O'Riordan, 1981). The contributors to Part One also confirm Sandbach's (1980) view that the milieu in which the nuclear industry and its opponents operate is far from being a pluralist democracy. These impressions become firmer in Part Two of the book. Whereas Part One particularly illuminates the element of accountability and control in the nuclear conflict, and its international dimension, Part Two highlights the changing political climate in which the conflict occurs, and in which the struggle for legitimacy is taking place. It illustrates the growing challenge posed by the anti-nuclear movement. This challenge has played on what we have called the Achilles Heel of the industry; its earlier neglect of the back end of the cycle. Blowers and Lowry (Chapter 6) Resnikoff (Chapter 8) and Pasqualetti (Chapter 9) all portray a remarkable situation, which indeed smacks of the uncertainty, prevarication and tendency to error of technocentrism: there appears to have been virtually no discussion of how to deal with the back end of the cycle, let alone any planning for it. In the view of Stewart and Prichard of the US Nuclear Regulatory Commission (Chapter 7), the important debate over nuclear waste is largely over where to dispose of it. But, as with nuclear power plants, to locate the debate thus, in siting questions, could mean that overall strategic planning and policy issues are neglected.

For nuclear waste, Blowers and Lowry and Resnikoff show that where you put it is but one of the issues - whether you continue to generate it at all (by continuing to have a nuclear industry), what you do with it (storing or reprocessing), and how you classify it (high or low level) all must be discussed. After reading Part Two, one is particularly impressed by the force of O'Riordan's later comment, that the hegemony of the nuclear industry's scientists and technicians is now under attack from 'non-establishment' science. In the dispute about waste, non-nuclear-industry scientists are challenging the nuclear scientists on their own ground, in technical matters. Both sides clearly feel that they must use science in the struggle for legitimacy.

In questioning the industry's back-end plans, or lack of them, more so than over front-end issues, the anti-nuclear movement seems to have struck a chord with public opinion, which is very suspicious of, and concerned about, waste disposal. In Britain the Nuclear Industry Radioactive Waste Executive (NIREX) has gained little credibility by its assurances to the people - partly, perhaps, because its claim to be exercising open-minded impartiality and objectivity may seem <u>in</u>credible since its members have such clear links with the nuclear industry. Part of the hallmark of the nuclear state is the way in which its controlling bodies and its watchdogs are drawn from the same <u>coterie</u> as that which actually makes up the industry. One could correctly claim this is unsurprising and unavoidable - where else could nuclear experts come from apart from the nuclear industry - (and the universities)? Of course this point may be conceded, but then so also must it be recognised that subsequent bias in the controlling and regulatory bodies - in favour of the industry - would also be very difficult to avoid. There are dissenting nuclear scientists who dislike aspects of the industry's activities, but they must always oppose from the outside, which means that they usually must leave the industry. They have formed a valuable reservoir of nuclear expertise from which environmental groups can draw.

There are two components of the public antipathy towards the nuclear industry. One is a growing awareness of what has happened; of mistakes made. Three Mile Island or the Sellafield discharges are, of course, spectacular examples, but the kind of errors which Resnikoff describes are perhaps more numerous and insidious. The other is the perception of what <u>might</u> happen - a fear of danger from radioactive wastes or decommissioned power stations: a fear that the nuclear state will ultimately have too much power over the majority of people; a fear that it does not anyway know what it is doing. It is this component which Stewart and Prichard identify as the greatest threat to the industry's back-end activities. Indeed, the whole industry could be permanently arrested in the US by public opposition: expressed not so much about broad nuclear issues, but on a site-by-site basis by local groups, coalitions and alliances. It seems, from Stewart and Prichard's analysis, that communities which are not already familiar with the front end of the cycle are unlikely to accept the imposition of back-end activities. But where there already has been mining and milling, nuclear power generation and weapons manufacture, then people are prepared to identify with the nuclear industry, and to see it as an indispensable part of the ethos of economic growth which they fervently support. Thus in remote, small communities, when political conditions are propitious, nuclear facilities may be welcomed. Bearing in mind, however, that conflict over nuclear issues is in a state of transition, we should consider the possibility that even 'pro-nuclear' areas may become less fervent in their support for the industry.

The nuclear state and the local community

Part Three of this volume focuses on local community impacts. Again, legitimacy, accountability and control, and the changing political contexts are elements which come to light. All three chapters suggest that it is perception rather than any 'objectively assessed reality' which governs people's responses and actions. Macgill and Phipps's survey of attitudes in the area around Sellafield (Chapter 10) confirms Stewart and Prichard's view that those who work for or live in the environs of the nuclear industry are less likely to regard present activities as hazardous, and are more likely to accept extensions of these activities. Indeed, the Yorkshire TV programme which in 1983 highlighted the cancer clusters around Sellafield was the subject of bitter complaints by many locals: there was no question of their gratitude for enlightenment about potential hazards. At one level, Macgill and Phipps's findings are obvious: where jobs are on the line workers will clearly be vigorous to defend their interests - their bliss will lie in ignorance about the nuclear industry's failings. Yet this is a point which, it seems, the anti-nuclear movement in Britain, like most environmentalists, often fails to grasp. If the point were to be grasped, then environmentalists would not only try to expose the nuclear industry's problems, but would also develop in some detail realistic solutions, such as 'alternative' production (TGWU, 1982), which can gain public and political support.

Macgill and Phipps also show that there is rising concern about cancer risk to children: even in the local Sellafield area, confidence in the nuclear industry is shaken. Those who are worried, again surprisingly, are fairly oblivious to warnings of the need to be cautious in appraising the true magnitude of the hazard. Hence the Black Report, suggesting that although there is a significant cluster of childhood cancers around Sellafield no significant causal link between them and the nuclear industry has been established, was roundly and swiftly condemned as a 'white-wash' by organisations like Greenpeace. Yet, as we can see from Craft and Openshaw's research (Chapter 11), it is quite correct that on any scientific basis we cannot show yet such a causal link. Other significantly high clusters exist elsewhere, and some of these are not near any nuclear installations. Zeigler and Johnson's chapter (12) illustrates how public response to any actual hazard is unlikely to be precisely tempered to the magnitude of the hazard. Neither 'objective' data about the Three Mile Island accident, nor public reassurances from the authorities, dissuaded many thousands of people in the vicinity of the plant from evacuating far further from it than had been officially decreed to be necessary. (Again there was a differential effect, whereby those who worked for the industry reacted less extremely than those with a history of anti-nuclear feelings.) This chapter shows that in planning for the continuance and expansion of nuclear activities it is this perceived rather than 'real' risk environment which must be catered for. The extent to which the evacuation issue has not been thought out by government officials was highlighted when Zeigler gave evidence at the Sizewell Inquiry.

Researching the nuclear state

This book confirms the view that there is a lack of coherent planning for energy in general and for all the activities associated with nuclear energy in particular, which is common to several Western countries. What planning there is, is generally confined by secretive processes rather than being open to democratic public consultation and involvement. It tends, too, to be restricted to specific siting matters and inquiries, although these should not and cannot be divorced from overall strategy. Furthermore, what consultation there is, is open to interpretation as a procedure for gaining retrospective acceptability for what may be a *fait accompli*. Like scientific research, planning is a far from neutral agent in such matters.

The conclusions in this book may well be confirmed by research into other areas of the nuclear cycle not covered here: for example, in mining, milling and reprocessing. Ad hoc incremental planning; an unbalanced political process of decision making; and a rising tide of public concern as the secrets of what has been done are revealed. All of these, we feel, may be characteristic of those parts of the nuclear cycle too. Further research is also needed into the question of transnational effects, already referred to here and examined in part by Pringle and Spigelman (1983), who show how remarkable similarities have occurred in the rise of the nuclear industry in many different countries. Other transnational issues which have been less well treated concern, for example, the export and import of nuclear technology, and the role of planning processes in it. For example, is the Sizewell Inquiry, as has sometimes been suggested, mainly to be seen as a matter of obtaining retrospective legitimation for the PWR, so that with a British clean bill of health it can be sold to China and the Far East? Can France's nuclear industry survive without its exporting technology to other countries? (Already, because of overproduction, electricity exports to Britain and other neighbouring countries may be essential to keep the French industry viable.) And to what extent is nuclear technology development within any one nation governed by the exigencies of international exchanges of material and technology for nuclear weapons production?

This last question leads us on to a still more neglected area - the connection between nuclear power and nuclear weapons. So many of the themes and problems discussed in this book may be more related to this connection than any of our authors are able to say. Few of them would be likely to assert that the two can be meaningfully divorced, but showing the extent of the marriage is difficult. It is in this key area that the closed and secretive nature of the nuclear state is most obvious. Opening it up to public scrutiny and critical appraisal is surely one of the most important tasks which academics, like the contributors to this book, have to carry out in future.

REFERENCES

ALBURY, D. and SCHWARTZ, J. (1982) Partial Progress, the Politics of Science and Technology, London: Pluto Press.

BOOKCHIN, M. (1980) Towards an Ecological Society, Montreal: Black Rose Books.

COTGROVE, S. (1982) Catastrophe or Cornucopia: the Environment, Politics and the Future, Chichester: Wiley.

DEPARTMENT OF THE ENVIRONMENT (1984) An Incident Leading to the Contamination of the Beaches near the British Nuclear Fuels Ltd. Windscale and Calder Hall Works, Sellafield, November 1983, Radiochemical Inspectorate, February.

EDWARDS, R. (1985) Nuclear Power, Nuclear Weapons: the Deadly Connection, CND Publications.

GUARDIAN, 27 December 1984.

GUARDIAN, 8 & 9 May 1986.

GUARDIAN, 8 July 1985.

GUDGIN, G. and FOTHERGILL, S. (1984) The Economic Consequences of the Sizewell 'B' Nuclear Power Station, Cambridge: Dept. of Applied Economics.

HEALTH AND SAFETY EXECUTIVE (1984) The Contamination of the Beach Incident at British Nuclear Fuels Ltd., Sellafield, November 1983, February.

HMSO (1976) Nuclear Power and the Environment, Royal Commission on Environmental Pollution, Sixth Report, London: HMSO (Command 6618).

HMSO (1984a) Investigation of the Possible Increased Incidence of Cancer in West Cumbria: Report of the Independent Advisory Group, Chairman; Sir Douglas Black (The Black Report), July, London: HMSO.

HMSO (1984b) Report of the Independent Review of Disposal of Radioactive Waste in the North East Atlantic, Chairman; Professor Fred Holliday (The Holliday Report), December, London: HMSO.

HMSO (1986A) Radioactive Waste, First Report of the Environment Committee 1985-6, House of Commons, March, London: HMSO.

HMSO (1986b) 'The assessment of best practicable environmental options (BPEOs) for management of low and intermediate solid radioactive wastes', Radioactive Waste (Professional) Division of the Department of the Environment, March, London: HMSO.

O'RIORDAN, T. (1981) Environmentalism, London:

Pion, Second Edition.
PORRITT, J. (1984) Seeing Green, Oxford: Blackwells.
PRINGLE, P. and SPIGELMAN, J. (1983) The Nuclear Barons, London: Sphere.
SANDBACH, F. (1980) Environment, Ideology and Policy, Oxford: Blackwells.
SCHNAIBERG, A. (1980) The Environment: From Surplus to Scarcity, New York: Oxford University Press.
SCHUMACHER, E. F. (1973) Small is Beautiful: Economics as if People Really Mattered, London: Abacus.
SCHUMACHER, E. F. (1980) Good Work, London: Abacus.
TGWU (Transport and General Workers' Union) (1983) A Better Future for Defence Jobs, London: TGWU.

Chapter Two

INTER-REGIONAL FACTORS IN THE LOCATION OF NEW NUCLEAR POWER STATIONS IN ENGLAND AND WALES

John Bradbeer

The electricity supply industry in England and Wales hopes, in the mid-1990s, to embark upon a new programme of nuclear power station construction. This chapter sets out to analyse at the international level some of the main issues which this programme will raise. The industry's own perspective of total systems cost is adopted, and this shows the range and complexity of locational choice. However, very considerable uncertainty exists in the economic, social and political environment within which the industry operates and future locational choice is far from clear cut.

In planning for the future, the electricity supply industry must consider several factors, each with some degree of uncertainty. The industry's stock of generating plant of different types is known and the pattern of future availability may be calculated fairly accurately from likely retirement and plant currently under construction. Much less certain is the price of fuel in future, an uncertainty which fairly recent experience with oil prices has underlined. Also less certain is the demand for electricity, which has fluctuated considerably in the last fifteen years. The industry can exercise a limited influence on demand by tariff changes but demand elasticities themselves change through time. Other important factors are quite beyond the industry's control. Energy policy in the United Kingdom has usually been expressed in general terms and frequently has been overtaken by international events. The industry knows that its nuclear programme enjoys Conservative government support but the position of the Opposition parties, should they come to power, is far less clear. Whether nuclear electricity is seen as replacing some of the non-electricity energy demand is unclear, and with there being no national energy policy, it is a major area of uncertainty. Public hostility to nuclear power seems fairly widespread but whether it will be sustained and what the response of government will be should this happen is also uncertain.

A final element of uncertainty concerns the possible privatisation of the electricity supply industry. The conclusions reached here could be altered if major organisational changes were to accompany privatisation. If the option chosen involves regional, all-purpose generating and supply utilities, on the West German or United States models, then each utility would seek independently to minimise its own system costs. Many would therefore seek to control their own supplies by building generating plant and the lure of the prestige, if not of the economics, of nuclear plants might be very strong.

This chapter does not attempt to examine all the elements of uncertainty facing the industry, but rather adopts as a starting point the likely consequences of decisions already made and trends and developments already becoming evident. The analysis commences by examining locational choice, and issues in total systems costs. It will be shown that the locational decision is but one of many facing the industry, and that it is usually regarded as a relatively residual independent variable. The total systems cost perspective also emphasises the interdependence of generation, distribution and demand, and the wide-ranging impact of the introduction of new generating plant. The demand for electricity is examined at the regional scale, from which a more accurate assessment of future demand can be made and the range of locational choice more clearly defined. In particular, current trends show clearly that there are regions of expanding electricity demand at the same time that other regions have stagnating demand. The relatively more certain supply side is also examined at the regional scale, and the present and shorter term future availability of generating capacity is evaluated. Some regions appear both now and in the future as major areas of generating capacity, while others have very little but may be seen as target areas for new power station locations. Supply and demand balances at the regional level are analysed, and a clear geographical pattern emerges of regions of major over-supply and of deficit. New nuclear power stations will be introduced into such a spatially-differentiated system. A major constraint upon locational freedom is the capacity of the distribution network, which will be under considerable pressure if present trends continue. The chapter concludes by examining three possible patterns of locational choice for new nuclear power stations, subject to the constraints of inter-regional electricity balances and the capacity of the present distribution network.

Locational studies of the electricity industry and total systems cost approaches

Until the work of Hauser (1971), geographical analyses of power station location either were site descriptive or followed neo-classical economic principles. Nuclear power station location studies have emphasised the specific site requirements of such plants (e.g. Mounfield, 1961, 1967; Mason, 1971; Openshaw, 1980, 1982a, 1982b, 1984) and have pointed out that, as they did not require primary energy inputs, they were relatively locationally flexible at the inter-regional scale. The major principles in the location of fossil fuel fired plants were analysed by Manners (1962). He identified the problem at the inter-regional scale as that of choosing whether to move electricity or the primary fuel to the market. From examples in the United States and the United Kingdom, he showed that it was cheaper to move large quantities of electricity over long distances than to move the primary energy required to generate it. This was reinforced when considering the division of electricity demand into base load, which is ever present, and peak load, which is of short duration. Base load generation enables full economies of scale to be realised in both production and distribution. It is therefore located at the cheapest sources of primary energy. Peak load generation, by contrast, is market orientated because short duration of demand makes electricity transmission over any distance uneconomic. Nuclear power stations, freed from a dependence on energy inputs, have a greater degree of locational flexibility than other stations. They may be used for base load generation anywhere there is a sufficient level of demand and would be especially attractive in regions of high demand but lacking cheap fossil fuels.

Hauser (1971), drawing on work by power engineers and influenced by studies by industrial economists such as Turvey (1963, 1986), introduced the concept of total systems costs. This recognises that the optimum state of the industry is attained by a simultaneous manipulation of generation, demand and distribution, with location not as a primary but as a relatively minor variable. The industry tries to plan an optimal pattern of generation and distribution to meet demand in each of the 8760 hours in the year. Longer term planning equally involves considering the whole system and impacts of change in any one component upon the others.

A major tool in planning and organising the supply side is the merit order, a ranking of power stations by operating costs. Modern plant has lower operating costs than older plant and it thus occupies the top positions in the merit order. Hydro-electricity stations, of which there are few, have low operating costs irrespective of age. This is also true of nuclear power stations, although to sustain low operating costs many have been derated; that is their capacity has been reduced. Oil-fired power stations have suffered until recently from soaring fuel prices since their construction and they have never occupied the high merit order places intended for them, except during the miners' dispute. Plants at the top of the merit order are allocated to base load generation and plants lowest in the merit order generate only at time of peak demand. Plants in the middle of the merit order are used to generate the day-time base load throughout the year and will typically be used to about 40-50 per cent of annual capacity. As new plant enters the merit order, it will naturally displace older and less efficient plant and thus reduce generation costs in total. New plant has incurred substantial capital cost in construction and intensive use is desirable to allow a quick return on investment. In a system where demand is static, too rapid introduction of new plant will mean premature relegation in the merit order of plants which have yet to be fully depreciated, thus raising rather than reducing total cost.

Until recently, most low merit order plants have been old, fully depreciated, relatively small and fossil fuel fired. As they have been retired, some new plants have been built specifically to meet peak demand. They have been gas turbine plants; cheap to build but up to three times as expensive to operate as modern base load stations.

Gas turbine plants offer two advantages over older but larger fossil fuel plants now relegated in the merit order. First, as they are smaller it is possible more exactly to match their total capacity to demand peaks and, secondly, gas turbines can be started up to give maximum output within minutes, but conventional plants require several hours to run up to the desired output. This could become significant in the future, as progressively larger coal-fired plants, which are however unsuited to peak load operations, become relegated in the merit order. This is because of the great diseconomies of scale incurred by operating plant at a fraction of its potential capacity.

New nuclear power stations will occupy positions high in the merit order and will displace quite large coal-fired plants. While most of the very large coal-fired plants located near the lowest-cost coalfields in South Yorkshire and the East Midlands would continue to be used for base load generation, other quite large plants would be relegated to peak load generation, with the problems that this entails. The smallest and oldest coal and oil-fired stations around the country would be retired completely, a process already under way. Nuclear power stations could be introduced to fulfil two functions. First, they could be required to meet incremental growth in electricity demand and would have relatively little impact upon the use of other base load stations. However, as will be shown later, the demand for electricity in England and Wales is not growing appreciably. Secondly, nuclear power stations could be required as replacement capacity for older coal-fired plants. Changes in the merit order and in the use of power stations would clearly follow from this and would also alter the flows of electricity through the distribution grid. The grid is operated so as to be used most effectively in each of the hourly planning periods in the year. Major alterations in the desired flows of electricity may be constrained by the capacity of the grid and could become a significant element in the choice of new power station locations.

Electricity demand

Demand for electricity in England and Wales has grown substantially since nationalisation and it is now five times that of 1948/9. During the 1950s it more than doubled and between 1960/1 and 1972/3, the average annual growth rate was 9.7 per cent. Since then, growth has been more restrained, although a peak was reached in 1978/9, and sales of electricity in the 1980s have been little above those of the mid 1960s. The pattern of sales by Area Electricity Board (AEB) is shown in Table 2.1. These figures are of recorded demand and are therefore net of distribution losses, internal consumption by the industry, and the growing but illegal practice of consumers by-passing their meters.

Also shown in Table 2.1 is the composition of electricity sales in each AEB by three major classes of consumer. Other sales are to farms and to traction and lighting authorities. Domestic consumption is characterised by marked peaking of demand, while industrial consumption is the major element of the base load. Commercial demand makes some contribution to base load but essentially comprises the major block of day base load and is subject to seasonal peaking. It can be seen that there are considerable variations in the sectoral pattern of AEB sales. London is most distinctive, with the largest share of commercial sales, and generally domestic demand is more important in the South and industrial demand in the North.

Table 2.2 shows recent patterns of change for the major sectors of consumption in each AEB. It can be seen that London, Southern, South Western and Eastern had greater sales in 1983/4 than in the national peak year of 1978/9 and that the latter three AEBs uniquely saw a growth in industrial sales. Domestic sales declined everywhere but commercial sales rose, although more rapidly in the South than in the North. The effect of these changes has been to reduce slightly the national base load but also to cut back on the peaking of demand. In the South the base load has grown but not by as much as the day time base load. In the North, the system has seen all types of demand depressed. In practice, this growth in the average load factor, that is the average demand expressed as a percentage of the annual maximum peak demand, has enabled the industry more readily to minimise costs and move towards a more optimal use of the system.

TABLE 2.1 SALES OF ELECTRICITY IN ENGLAND AND WALES

Area Electricity Board	Unit sales GWh (Gigawatt hours) 1978/9	1983/4	% Composition 1983/4 Domestic	Industrial	Commercial
London	15058	15118	37.0	12.4	48.2
South Eastern	14118	13952	47.9	24.1	25.8
Southern	19792	20187	43.6	30.0	24.0
South Western	10206	10340	44.9	27.7	22.7
Eastern	22226	22846	44.5	26.8	25.8
East Midlands	18289	18142	34.5	44.3	18.4
Midlands	20888	18934	36.8	40.4	20.3
South Wales	10543	10200	24.8	58.7	13.8
Merseyside & North Wales	14585	14543	27.9	53.7	15.9
Yorkshire	21642	20339	29.0	53.2	15.6
North Eastern	13310	13003	30.3	48.6	18.0
North Western	18935	17966	35.5	39.8	22.1
Direct Sales by CEGB	5668	3944			
TOTAL	205260	199514			

Source: Data from Electricity Council Annual Reports

TABLE 2.2 CHANGES IN ELECTRICITY SALES 1978/9 - 1983/4

Area Electricity Board	Per cent change 1978/9 - 1983/4 Domestic	Industrial	Commercial	Total
London	-8.2	-18.3	+11.3	+0.4
South Eastern	-6.1	-5.3	+14.6	-1.2
Southern	-4.6	+4.2	+15.4	+2.0
South Western	-5.2	+2.3	+15.2	+1.3
Eastern	-3.5	+1.6	+18.5	+2.8
East Midlands	-0.2	-5.5	+13.4	-0.8
Midlands	-8.4	-17.5	+9.4	-9.4
South Wales	-5.3	-5.4	+9.6	-3.2
Merseyside & North Wales	-6.8	-0.6	+14.0	-0.3
Yorkshire	-7.2	-9.0	+9.1	-6.0
North Eastern	-4.2	-6.8	+13.9	-2.3
North Western	-8.1	-8.6	+7.8	-5.1
TOTAL	-5.5	-6.2	+12.8	-2.0

TABLE 2.3

GENERATING CAPACITY BY PLANT TYPE AND REGION - 31.3.84

Installed capacity, MW (Megawatts)

Area Electricity Board	COAL pre-1960	COAL post-1960	OIL pre-1960	OIL post-1960	NUCLEAR	HYDRO AND PUMPED STORAGE	GAS TURBINE	TOTAL
London	-	-	330	345	-	-	140	815
South Eastern	308	1920	-	2691	410	-	318	5647
Southern	-	1820	95	1932	-	-	240	4087
South Western	85	-	-	-	1470	3	-	1558
Eastern	347	2548	248	-	665	-	458	4266
East Midlands	1494	5948	-	-	-	-	338	7780
Midlands	1600	3636	-	-	700	-	314	6250
South Wales	338	2027	-	1900	-	-	151	4416
Merseyside & North Wales	228	1880	-	480	1230	999[a]	118	4935
Yorkshire	908	6917	-	-	-	-	367	8192
North Eastern	972	1100	-	-	-	-	68	2140
North Western	802	112	-	-	-	-	-	914
TOTAL	7082	27908	673	7348	4475	1002[a]	2512	51000

[a] includes 890 MW pumped storage

Source: Data from CEGB Statistical Yearbook 1983/4

TABLE 2.4

GENERATION BY PLANT TYPE AND REGION 1983/4 GWh (Gigawatt Hours)

Area Electricity Board	COAL pre-1960	COAL post-1960	OIL pre-1960	OIL post-1960	NUCLEAR	HYDRO	GAS TURBINE	TOTAL
London	-	-	14	182	-	-	9	205
South Eastern	826	7000	-	2115	2993	-	7	12941
Southern	-	10064	-	2007	-	-	8	12079
South Western	6	-	-	-	10663	10	-	10679
Eastern	279	7794	124	-	4221	-	14	12432
East Midlands	6420	42452	-	-	-	-	2	48874
Midlands	4746	21040	-	-	3749	-	9	29544
South Wales	243	10226	-	1985	-	-	-	12454
Merseyside & North Wales	576	10642	-	38	8984	191	0	20431
Yorkshire	2327	40664	-	-	-	-	3	42994
North Eastern	3407	3459	-	-	-	-	1	6867
North Western	2269	593	-	-	-	-	-	2862
TOTAL	21099	153934	138	6327	30610	201	53	212362

Source: Data from CEGB Statistical Yearbook 1983/4

Electricity generation and supply

Generating capacity has expanded five-fold since nationalisation. During the 1950s and early 1960s new capacity was added rapidly so that the average annual rate of growth was in excess of 7 per cent. Expansion was more modest in the 1970s, although a peak of 58,677 MW of capacity was reached in 1975/6. Since then capacity has been reduced by 13 per cent and further closures were announced by the Central Electricity Generating Board (CEGB) in 1982. The effect of this will be to reduce the plant margin, the excess of capacity over probable peak demands, to levels more appropriate to meet likely contingencies.

Table 3.3 shows the net capacity of the generating system by plant type and allocates it to the appropriate AEB. Coal-fired plants account for over two-thirds of net capacity, and oil-fired plants for a further sixth. Nuclear plants represent about 9 per cent of net capacity but pre-1960 coal-fired plants actually account for a greater share. Each AEB contains some generating capacity, but relatively little is located either in London or in North Western.

The objectives of minimising total costs yield a spatial and plant type pattern of generation shown in Table 2.4. This contrasts with the spatial and plant type pattern of generating capacity shown in Table 2.3 in a number of ways. Coal-fired plants accounted for over 80 per cent of the actual generation and oil-fired plants for only 3 per cent. Generation by nuclear plants accounted for just over 14 per cent, considerably more than their share of capacity. Major contrasts can also be seen between AEBs. East Midlands, Midlands and Yorkshire supply over half of the electricity and there is relatively little generation in the South and East of the country. This reflects the location of coal-fired base load stations and the presence in the South of major oil-fired base stations, now only used for generating daytime load, and older stations relegated to peak load generation. Sudden surges of demand are met by gas turbine, widely dispersed, but whose contribution to generation are small.

Generating plant currently in commission varies considerably in age and much is already 25 years old (Table 2.5). In the short term a quite considerable reduction in the plant margin will be obtained by retiring such plants, most of which are

TABLE 2.5 EXISTING GENERATING CAPACITY BY TYPE AND AGE, ENGLAND AND WALES - 31.3.84.

DATE Commissioned	MW Coal	Oil	Nuclear
1949-1959	7082	673	-
1960-1969	17207	3075	2605
1970-1979	10665	3162	1880
1980-1984	-	1111	-
Under construction	660	660	5070

Source: Data in CEGB Statistical Year Book 1983/4

small and very low in the merit order. By the early 1990s about half of the present coal-fired capacity will be this old and will have become less efficient and reliable. This will pose the industry new problems. Traditionally, older plant has had some years of peak load generation before being retired. For the reasons outlined earlier, large power stations of any type are ill-suited to this task and may well be retired completely. This would apply both to the Magnox stations of the first nuclear power programme and to coal-fired stations. Replacing large coal-fired stations raises another issue, for nuclear power stations have been on average only about 1,000 MW in capacity whereas most of the coal-fired plants have been of 1,500 MW or larger. Thus unless the capacity of individual nuclear power stations is greatly increased, three nuclear power stations will be needed in replacing every two coal (or oil) plants being retired. The number of new sites required will also increase, even allowing for continuation of current practice at several locations where two nuclear power stations are on adjacent sites.

Regional supply and demand balances

Taking the operational costs of different types of plant and their distribution between AEB regions, it is clear that a distinctive pattern of electricity supply will be revealed. A way of examining an idealised pattern of supply is to calculate a generation output for each AEB, assuming that each of its plants was to be used exactly as the national average for its type. This method uses 1983/4 load factors for different plant types. Thus London has 330 MW capacity in pre-1960 oil-fired stations, which nationally had a load factor of 2.3 per cent. With 8760 hours in the year, such plants in London could have generated 330 x 8760 x 2.3% MWh or 66.5 GWh, but were used to produce only 14 GWh (see Table 2.4). The potential generation output using this method is shown in column 1 of Table 2.6 as a percentage of actual demand in each AEB. The Table shows that six AEBs could meet more than their demand for electricity and that a seventh could come quite close. By contrast, London and North Western could only hope to meet a small proportion of their demand.

In practice, the actual operation of the system departs a little from this idealised pattern and is shown in column 2 of Table 2.6. The South Eastern region, which has the potential to be self-sufficient, is actually a net importer of electricity, and in Eastern and North Eastern regions generation is lower than the theoretical potential. Significantly more generation occurs in East Midlands and Yorkshire than would be expected. Column 3 of the Table shows actual generation in pre-1960 stations as a percentage of demand. This is quite significant in East Midlands, Midlands and North Eastern regions and of some consequence in Yorkshire and North Western regions. Retirement of all of these plants, which will be virtually completed by 1990, would still leave a considerable surplus of capacity over demand in East Midlands, Midlands and Yorkshire, but it would increase the need for electricity imports to North Eastern. London and North Western would be left with virtually no generating capacity.

New plant under construction will affect only three regions but add significantly to the generating potential of each. In the South Eastern region, net export potential will be increased, while in North Eastern and North Western regions, even after plant retirements, near self-sufficiency will result. Given the lengthy period of preparation, obtaining planning permission and construction which attends any new power station, it is appropriate to examine the distribution of plants which will be 25 years old in the mid 1990s. In fact half of current capacity will be of that age in 1995 and it is worth noting that seven nuclear stations are included. In five regions over half of the capaci ty will be 25 years old and in a further four, a third or more of capacity will be of this age. Only in the Merseyside and North Wales and North Western regions would no new capacity be required as replacement for plants retiring in the late 1990s.

The likely evolution of inter-regional and inter-sectoral demand constitute further considerations in future supply and demand balances. The net supply AEBs today (East Midlands, Midlands, South Wales, Merseyside and North Wales and Yorkshire), are those with a substantial industrial demand (see Table 2.1) but they are also those where industrial demand has <u>fallen</u> most sharply. Although continuing decline at recent rates is unlikely, the industrial demand for electricity in these regions will not regain former levels, reflecting the impact of industrial restructuring and improved efficiency in energy use. These same AEBs have been unable to expand sales to commercial consumers at anything like the national rate and, therefore, their potential supply surpluses have grown and probably will remain higher than before.

AEBs such as London, South Eastern, Southern, South Western and Eastern are major importers of electricity and are also those where demand has recovered quickly from the low levels of the early 1980s. In large, part of this has been because of a combination of buoyant sales to commercial consumers and growth in the industrial demand. The North Eastern and North Western regions are also importers of electricity, but they do not have especially dynamic electricity markets.

The continuation of all of these trends in the short-term seems probable. In the short-term, they can be accommodated, subject to constraints imposed by the distribution network. Growth in available export surpluses and declining base load in some AEBs complements well the growing deficits and increasing base load in many of the others. In the medium term, adaptation to this evolving pattern is possible and desirable.

The original national grid was built with a voltage of 132 kV and was expanded rapidly throughout the 1950s (Table 2.7). The capacity of this network to transmit large quantities of electricity was fairly limited and with improvement in power transmission technology, new trunk mains were built at 275 kV. The pace of technological change was such that by the mid 1960s a new standard of 400 kV was set. The technical and economic advantages of

TABLE 2.6

REGIONAL SUPPLY AND DEMAND BALANCES 1983/4

Area Electricity Board	Theoretical potential generation as % electricity purchases	Actual generation as % electricity purchases	Generation in pre-1960 stations as % electricity purchases	Potential output[a] of plant under construction as % electricity purchases	Plant Commissioned 1961-1970 as % total capacity
London	2.2	1.2	0.1	-	42.3
South Eastern	109.3	85.8	5.5	59.9	47.3
Southern	54.3	55.9	-	-	47.3
South Western	93.1	95.2	0.1	-	27.6
Eastern	79.8	50.7	1.6	-	74.5
East Midlands	205.8	253.8	33.4	-	72.1
Midlands	145.0	147.2	23.7	-	39.9
South Wales	125.3	114.8	2.2	-	50.3
Merseyside & North Wales	127.1	127.4	3.6	-	7.9
Yorkshire	188.8	201.5	10.9	-	56.3
North Eastern	67.8	50.2	24.9	63.4	51.4
North Western	12.6	14.9	11.8	87.1	12.3

[a] Assumed load factor for coal and nuclear plant 75% and for oil plant 25%

Table 2.7 MAIN TRANSMISSION NETWORK

circuit km

Voltage	400kV	275kV	132kV	Total
At year end				
1950/51	-	-	8629	8629
1955/56	-	1063	10293	11356
1960/61	-	4528	14687	19215
1965/66	570	6286	18397	25253
1970/71	6587	5316	19323	31226
1975/76	8477	4385	19245	32107
1980/81	9425	4422	19168	33015
1983/84	9531	4307	19113	32951

On 1.4.69 most of the 132kV network was transferred to AEBs. The higher voltage networks remain under CEGB control.

Source: Electricity Council Annual Reports

this standard were greatly enhanced by the fact that the existing 275 kV network could be up-rated to it quite readily and little new route-kilometerage was required. In the last 15 years the average annual growth of the route-kilometerage has been 1 per cent although that of 400 kV network has been three times as great. The grid system now is fairly close to its ultimate capacity and either greater flows of electricity or altered flow patterns will require it to be expanded. No further improvement of standards using the existing network is possible on the scale required, so the options open will either be to construct completely new mains networks or to duplicate existing routes. Obtaining permission for overhead transmission lines has been especially difficult in the past but relatively little construction has taken place since the growth in environmental awareness and the amenity lobby in the early 1970s. There is no reason to suppose that new power lines will be any more popular with amenity groups than motorways have been. The option of underground routing is prohibitively expensive and is only used as a very last resort for short stretches in the most sensitive areas.

Another consideration in future balances is the resilience of the system in the face of unexpected contingencies. The impact of the blizzards in Southern England in the early 1980s was considerable and it led to the failure of grid supply and extensive and lengthy black-outs. The industry came under considerable pressure from public opinion, Parliament and the Department of Energy, to take steps to avoid a repetition. In the medium term, two responses are possible. One, would be to strengthen further the distribution network, especially by providing another north-south link and by reinforcing supply routes to major deficit regions, such as the south coast, the south west and East Anglia. The other would be to locate new base load power stations in or close to these areas. Indeed, the South Western Electricity Consultative Council, representing consumers, has publicly stated its concern at the absence of generating capacity in much of the region and it strongly supports the construction of a nuclear power station in Devon or Cornwall, perhaps reflecting its chairman's background as Flag Officer Nuclear Submarines.

Possible strategies for new nuclear power stations

It is clear that considerable replacement capacity will be needed in the electricity supply industry in the mid- to late-1990s. Even to maintain the existing balance between coal and nuclear plant, some new nuclear power stations will be required. It appears to be both present Government and CEGB policy to have a reduced reliance upon coal-fired power stations and should this policy be continued, then some coal-fired stations will be replaced by nuclear. Although the CEGB strenuously denies any association between its coal and oil fired power stations and acid rain, its conversion to the environmentalists' viewpoint would almost certainly reinforce its nuclear enthusiasm, rather than lead it to fit sulphur dioxide reduction and control technology to existing power stations.

In attempting to evaluate the industry's response to problems of planning new regional supply-demand balances for the later 1990s, three strategies appear to be possible apart from a shift towards energy conservation.

The first would involve relatively little change in the inter-regional distribution of generating capacity. The advantages of this would be that existing power station sites could be re-used. This would ease problems of access to cooling water, so essential to all large power stations, and it would require only a limited amount of new transmission network to link stations to the national grid. However, new sites will be required for several reasons. For the reasons outlined earlier, probably three nuclear plants will be required for every two coal-fired plants and many existing power station sites would not meet current nuclear power station siting requirements on safety grounds. Some nuclear power stations will be among those to be replaced and the time consuming process of decommissioning will preclude their re-use in the short term.

Major problems would follow from a search for new nuclear power station sites on or close to the coalfields where the bulk of the retiring coal-fired plants are located. Specific site needs, such as sound rock foundations and access to cooling water, as well as the perceived need to be distant from major population concentrations, would make locational search in such areas difficult. It is also likely that public support for nuclear power stations would be less in coal mining areas, especially in the aftermath of the 1984/5 miners' dispute. Many of these issues are well illustrated by the controversy over Druridge Bay in Northumberland as a potential nuclear power station site. Perhaps the most telling reason why this minimum-change option is unlikely to be chosen is that it would require the strengthening of the national grid to facilitate growing north-south flows of electricity if present trends in inter-regional and inter-sectoral demand for electricity persist.

The second strategy would entail replacing some of the retiring coal-fired plants in the North with new coal-fired stations on the same sites, but locating much of the replacement capacity in the South, where demand is growing. Given that much of this growth is of base load, this strategy appears to be most attractive. Almost all of the new capacity to be located in the South would be nuclear, as the economics of moving coal long distances to base load stations from inland coal fields would be unfavourable. Indeed, imported coal would be as likely to be the competitor to nuclear power in the South. This strategy would minimise the need for a new inter-regional distribution network to be built, although numbers of spurs would be needed to link new stations to the grid and perhaps some local strengthening might also be needed. Nonetheless, the strategy has major drawbacks, not the least of which is general public concern about nuclear safety and the loss of amenity.

The third strategy is both the most radical and the least likely to find favour as it looks towards a complete restructuring of the economy and the concentration of population and economic activity in the South. In this case, all replacement capacity would be located in the South and coal-fired plants elswhere would be closed without replacement, until at least the early years of the next century. The distribution network could remain unaltered but gradually the flows would be reversed. This strategy avoids most of the problems of the first but magnifies and intensifies those of the second. A more realistic variant of this strategy would be the peripheralisation of electricity generation. A gradually reducing core of coal-fired plants would be located on and near the low cost South Yorkshire and Nottinghamshire coalfields, and perhaps later near the Vale of Belvoir coalfield. New nuclear plants would be built on the coastal periphery of England and Wales, both in the North and more extensively in the East, South and South West. A few new trunk mains would be required but the dispersed nature of protest over transmission lines rather than the concentration of protest at a specific power station site, would be felt by Government and the CEGB as easier to combat.

It appears that the public will be faced with the choice of accepting either new power stations in the South of the country and limited expansion of the national grid, or a substantial growth of the grid system to accommodate changing demand and a relatively static spatial distribution of supply. New nuclear power stations offer both advantages and disadvantages to the industry. They would seem on technical and economic grounds to be ideal to re-balance electricity supply and demand in England and Wales where the growth of demand is concentrated in the area with least present generation stations and least access to fossil fuel. The stringent siting requirements on engineering and safety grounds, however, make site selection difficult and very contentious in areas with strong amenity lobbies; while the national grid imposes its own constraints on locational freedom. The industry finally has to weigh up the likely outcome of many uncertain elements and consider whether, far from solving problems, nuclear power might create more and intractable difficulties.

REFERENCES

HAUSER, D. P. (1971) 'System costs and the location of new generating plant in England and Wales', Transactions of the Institute of British Geographers, 54, 101-121.

MANNERS, G. (1962) 'Some location principles of thermal electricity generation', Journal of Industrial Economics, 10, 218-230.

MASON, P. E. (1971) 'Some environmental considerations in the siting of nuclear power stations along the California coast', Geography, 56, 335-337.

MOUNFIELD, P. R. (1961) 'The location of nuclear power stations in the United Kingdom', Geography, 46, 139-155.

MOUNFIELD, P. R. (1967) 'Nuclear power in the United Kingdom: a new phase', Geography, 52, 310-316.

OPENSHAW, S. (1980) 'A geographic appraisal of nuclear reactor sites', Area, 12, 287-290.

OPENSHAW, S. (1982a) 'The siting of nuclear power stations and public safety in the United Kingdom', Regional Studies, 16, 183-198.

OPENSHAW, S. (1982b) 'The geography of reactor siting in the United Kingdom', Transactions of the Institute of British Geographers (New Series), 7, 150-162.

OPENSHAW, S. (1984) 'An evaluation of the safety characteristics of current and future reactor sites', The Statistician, 33, 133-142.

TURVEY, R. (1963) 'On investment choices in electricity generation', Oxford Economic Papers, 15, 278-286.

TURVEY, R. (1968) Optimal Pricing and Investment in Electricity Supply, London: George Allen and Unwin.

Chapter Three

NUCLEAR ENERGY IN FRANCE: A FORETASTE OF THE FUTURE?

Miriam J. Boyle and Mike E. Robinson

France is a powerful industrial nation with large energy demands but few indigenous resources. Since the oil crisis of 1973 successive governments have been committed to closing this gap through the development of a massive nuclear energy industry. In some ways they have been very successful, translating policies into projects with startling rapidity. They now lead the West in many areas of civilian nuclear technology and they are certainly the most vigorous prosecutors of reactor construction both at home and overseas. We begin this chapter, therefore, by a consideration of the circumstances of this progress. We outline the highly integrated administrative and industrial structure of nuclear energy production and we describe the expansion of the reactor programme and the steps that have been taken to strengthen French control over every other aspect of the fuel cycle.

Impressive though these achievements have been, however, they have been purchased at considerable cost, both financial and social. We conclude the first half of the chapter by considering some recent economic assessments that dull the gloss of the French industry's pride. Haste and ambition, they suggest, now threaten a serious and costly mismatch between energy supply and stagnating demand. In addition, the massive capitalisation of the programme may be starving industrial development in other fields and increasing the imbalance of French commerce without making significant inroads on imported energy requirements.

In the second half of the chapter we turn to the social and political dimensions of the nuclear industry's expansion. In particular, we consider the tactics which have been adopted to secure the compliance of the mass of ordinary French people and we offer some essentially personal observations on the anti-nuclear movement and its apparent impotence in the face of a highly centralised bureaucracy. But the development of the industry also has a significance that extends beyond domestic issues. In many ways it serves as an unfolding scenario for the nuclear aspirations of other countries. Insofar as the French experience proves successful, therefore, it may signal the directions of energy investment elsewhere. On the other hand, if it serves principally to reveal the difficulties and dangers of a nuclear society, it may offer a cautionary message in policy development. Bearing this in mind, the final section of the chapter compares some general features of the relative progress of the British and French industries and suggests that there are circumstances which sometimes favour the tortoise rather than the hare.

Energy policy

Until 1973, the French probably envisaged a future tied largely to oil-fired electricity production. Sixty per cent of total energy demand was met with imported oil and the remainder was derived largely from coal and hydro-electric power. Nuclear investment was modest and conservative. Nine gas-graphite reactors of French design were constructed between 1952 and 1972 (Table 3.1), and a series of six pressurised water reactors (PWRs), built under licence from Westinghouse, were under construction at Fessenheim and Bugey (Table 3.3 and Figure 3.1). There is no evidence, however, to suggest a long-term strategy which would hinge critically on further nuclear developments. On the contrary, energy preoccupations were with oil and the consolidation of French overseas oil interests (Lucas, 1979).

The critical turning point came with the Middle Eastern conflicts of 1973 and with their impact on oil supplies. Five-fold rises in oil prices had a jarring effect on the economy and the relative impotence of the Government and the French oil industry severely dented national pride. The exposure of such vulnerability led to a resolve to minimise dependence on oil and to maximise the contribution of 'new' energy technologies. At the heart of this strategy was a determination to ensure French control over energy resources and consequently to guarantee their security. The key to the policy lay in a massive expansion of the nuclear programme and the establishment of national control over all aspects of the nuclear fuel cycle.

TABLE 3.1

FRANCE'S EARLY NUCLEAR REACTORS

Site	Reactor Type	Output (Mw.)	Opening Date
Marcoule G1	UNGG	2	1956*
Marcoule G2	UNGG	38	1959*
Marcoule G3	UNGG	38	1960
Chinon A1	UNGG	70	1963*
Chinon A2	UNGG	210	1965*
Chinon A3	UNGG	480	1967
Monts d'Arrée	HWR	70	1967*
Chooz A1	PWR	300	1967
St. Laurent A1	UNGG	480	1969
St. Laurent A2	UNGG	515	1971
Bugey 1	UNGG	540	1972

Notes * denotes reactors now closed
UNGG - Gas-graphite reactor
HWR - Heavy water reactor
PWR - Pressurised water reactor

Source: EDF (1982a); WISE (1984)

Organisation and decision-making

In a little over a decade the face of the French energy industry has been totally transformed. Elsewhere in the Western world there has been a hiatus in nuclear power development owing to a slackening in demand for electricity and to the uncertainties generated by the near-catastrophe at Three Mile Island. The French have deliberately exploited this lack of momentum to establish a position among the leaders in the league table of the nuclear nations (Table 3.2). At the heart of their achievement is an organisational and decision-making structure, dominated by an administrative elite, which co-ordinates and integrates the many different facets of the nuclear industry (Lucas, 1979; Pringle and Spigelman, 1983; Sweet, 1981; Ardagh, 1982).

TABLE 3.2

THE WORLD'S CHIEF NUCLEAR ELECTRICITY PRODUCERS

Country	Nuclear Power Production Capacity		Increase or Decrease of World Share (%) 1980-1985
	End of 1980 World Share (%)	End of 1985 Predicted World Share (%)	
USA	40.1	34.8	-5.3
France	9.7	13.0	+3.3
USSR	8.1	10.1	+2.0
Japan	10.9	7.1	-3.8
West Germany	6.5	6.2	-0.3
UK	6.2	4.2	-2.0
Canada	4.1	3.6	-0.5
Sweden	3.5	3.7	+0.2
Spain	0.7	2.6	+1.9
Belgium	1.3	1.9	-0.6
EEC	25.0	26.0	+1.0
World	100.0	100.0	--

Source: EDF (1982a)

As it stands today the organisation of the nuclear industry represents the outcome of a power struggle between two public utilities, the Commissariat à l'Energie Atomique (CEA) and Electricité de France (EDF). Until 1969, nuclear issues, civil and military, were the province of CEA. They were responsible for the design of the domestic gas-cooled reactors, called Uranium Naturel-Graphite-Gaz (UNGG), for the direction of research and development, and for the relationships of the nuclear sector with other sectors of French industry and the armed forces. The role of EDF was both more general and more passive. It was responsible for commissioning power stations and for generating, transmitting, and selling electricity. According to Pringle and Spigelman (1983), it had won universal respect for its control of electricity costs and its 'large and successful investment programme'. In 1970, however, EDF adopted a new commercial initiative (Lucas, 1979). Under the influence of its director-general, Marcel Boiteux, a new programme was designed to encourage an increase in electricity consumption. As part of this new and aggressive policy of expansion, Boiteux sought the replacement of indigenous gas-cooled reactors by PWRs of American design, ostensibly on the grounds of cost and efficiency. Inevitably the ambitions of EDF created a direct conflict with the vested interests of CEA: the former seeking to import a new reactor technology; the latter seeking to retain a domestic technology which they had developed and which they controlled. Moreover, since the two utilities were staffed from different 'Schools' ('Corps') of the French technocratic elite, the immediate dispute was exacerbated by deep-seated traditional rivalries.

The political manoeuvrings of CEA and EDF over the late 1960s were highly involved (Pringle and Spigelman, 1983). The eventual outcome, though, was to redefine the responsibilities of each organisation. EDF won the principal battle and assumed control of the reactor programme. As Lucas (1979) has observed, it has since become the driving force in the French nuclear initiative. Stripped of its earlier power, CEA, under the lead of André Giraud, recovered from its initial demoralisation and redirected its energies into other aspects of the nuclear fuel cycle and into breeder reactor technology. On the eve of the oil crisis, therefore, France had already engineered an administrative structure hinging on two organisations which were now complementary rather than competitive. The new directions taken by national energy policy after 1973 gave scope to both of them to enhance their institutional stature and to extend the industrial and commercial bases of their nuclear operations. In large measure they did this through private companies. A monopoly over building and installing reactors was given to the French company, Framatome, which enjoyed the patronage of Giscard d'Estaing and in which CEA secured a thirty per cent holding (Pringle and Spigelman, 1983). Other areas of fuel cycle activity were consolidated by CEA with a private subsidiary called the Companie Générale des Matières Nucléaires (Cogema). It is within this tight, clearly defined, and mutually-supportive structure that the French nuclear industry has prospered and grown.

TABLE 3.3

FRANCE'S 900 Mw. PWR PROGRAMME

Sites operating in 1984	Opening Date
Fessenheim 1	1977
Fessenheim 2	1978
Bugey 2	1979
Bugey 3	1979
Bugey 4	1979
Bugey 5	1980
Tricastin 1	1980
Tricastin 2	1980
Tricastin 3	1981
Tricastin 4	1981
Gravelines B1	1980
Gravelines B2	1980
Gravelines B3	1981
Gravelines B4	1981
Dampierre 1	1980
Dampierre 2	1981
Dampierre 3	1981
Dampierre 4	1981
Le Blayais 1	1981
Le Blayais 2	1982
Le Blayais 3	1983
Le Blayais 4	1983
St. Laurent B1	1982
St. Laurent B2	1982
Chinon B1	1982
Chinon B2	1983
Cruas 1	1983
Cruas 2	1984
Cruas 3	1984

Sites under construction in 1984	Planned Opening Date
Gravelines C5	1985
Gravelines C6	1985
Chinon B3	1986
Chinon B4	1986
Cruas 4	1985

Source: EDF (1982a); CEA (1983)

The reactor programme

The first stage in the pursuit of a new nuclear strategy involved cancelling all orders for oil-fired power stations and an immediate expansion of investment in PWRs (EDF, 1982b). In the medium term this could be achieved only by continued dependence on American technology, and Framatome was the French licensee for Westinghouse. By 1984 it had constructed a series of twenty-nine 900Mw. reactors and a further five, also built under Westinghouse licence, are scheduled for completion before 1990 (Table 3.3; Figure 3.1).

Despite the Westinghouse connection, however, the French have made no secret of their distaste for US domination of post-war nuclear developments. Westinghouse technology was exploited in order to permit acceleration of French experience and expansion in the industry's infrastructure without waiting for the outcome of domestic research and development on light water reactors. Even so, it has been crucial to longer-term strategy that the process of 'Franchification' should centre on the eventual switch to an independent technology. By 1980 this technology had progressed sufficiently to allow the licence agreement with Westinghouse to lapse and to substitute for the 900 Mw. reactors a new series of 1,300 Mw. reactors of indigenous French design (L'Express, 9th February, 1980; Samuel, 1982). The first of these, at Paluel on the Channel coast, is already connected to the grid, and a further fourteen units are in the planning or site-preparation stage (Table 3.4; Figure 3.1). In addition, the fast reactor programme is poised to begin commercial operation with the opening of the Superphénix plant at Creys-Malville near Lyon which was connected to the grid in January 1986 (Gallacher, 1984; O'Dy, 1984a).

It is clear that in exploiting an imported technology the French, with the co-operation of Westinghouse, have bought sufficient time to bring the scientific, industrial, and commercial bases of their own nuclear industry to maturity. In theory at least, seventy per cent of all French electricity and almost thirty per cent of total energy production will be supplied from nuclear reactors by 1990 (Boyle and Robinson, 1981a; Table 3.5). In relative terms this dwarfs the nuclear ambitions of other countries and even in absolute terms only the USA produces more electricity from reactors (Table 3.2).

Figure 3.1 FRENCH NUCLEAR REACTORS, CURRENT AND PLANNED

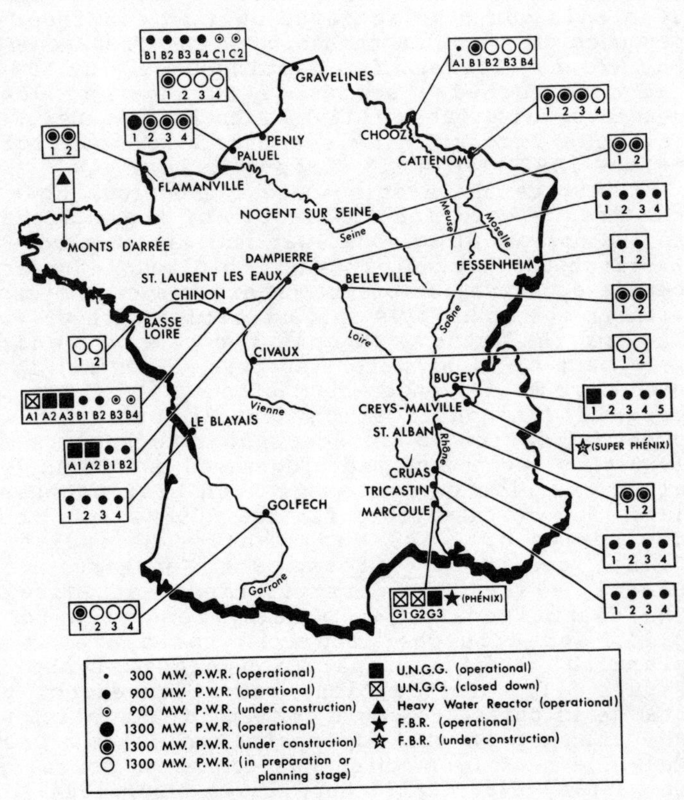

SOURCES - C.E.A. (1983) Quelques Informations Utiles
E.D.F. (1984) Various sources

Figure 3.2 THE NUCLEAR FUEL CYCLE

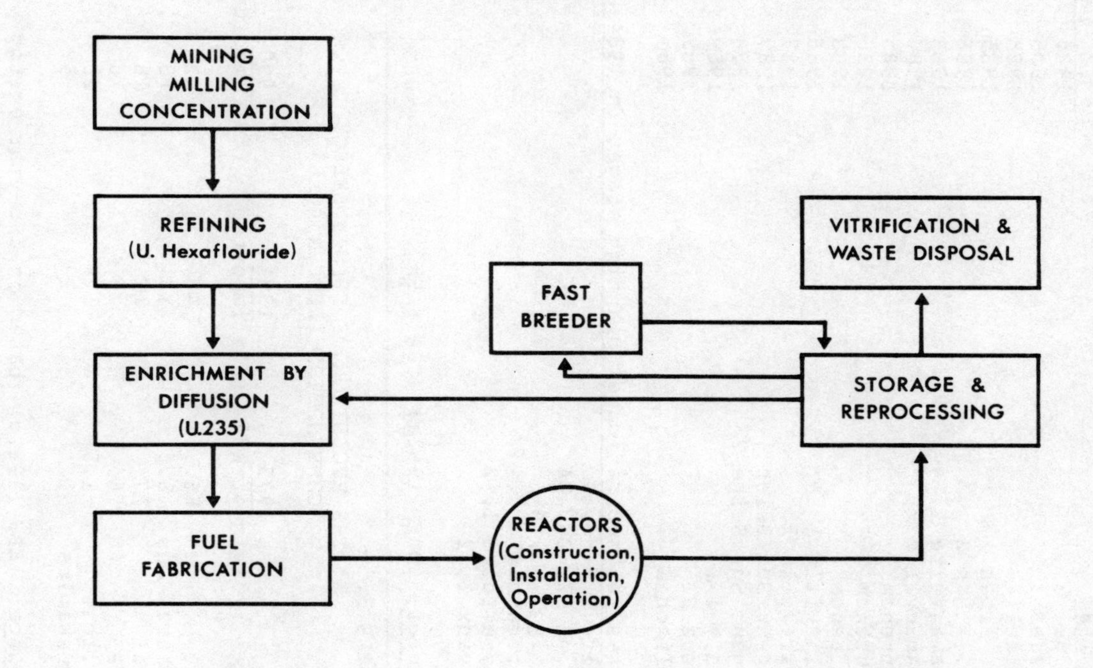

TABLE 3.4 FRANCE'S 1,300 Mw. PWR PROGRAMME

Sites in operation in 1984	Opening Date
Paluel 1	1983

Sites under construction in 1984	Planned Opening Date
Paluel 2	1985
Paluel 3	1985
Paluel 4	1985
St. Alban 1	1985
St. Alban 2	1986
Flamanville 1	1985
Flamanville 2	1986
Cattenom 1	1985
Cattenom 2	1986
Cattenom 3	1988
Belleville 1	1986
Belleville 2	1987
Nogent Sur Seine 1	1987
Nogent Sur Seine 2	1988
Penly 1	1989
Chooz B1	1989
Golfech 1	1989

Sites in preparation or planning stages in 1984

Golfech 2, 3, 4
Chooz B2, B3, B4
Penly 2, 3, 4
Cattenom 4
Basse Loire 1, 2
Civaux 1, 2

Source: EDF (1982a); CEA (1983)

TABLE 3.5

THE EVOLUTION OF FRENCH ENERGY CONSUMPTION (PERCENTAGES)

	1973 (75% imported)	1981 (65% imported)	1990 (50% imported)
Oil	66	48	32
Natural Gas	8.5	13	13
Coal	17	17	16
HEP	6	8	6
Nuclear	1.5	12	28
Renewables	1	2	5

Source: EDF (1982c) (based on two-year energy plan 1982-4, November 1981)

Fuel cycle operations

Electricity production from reactors depends upon a fuel cycle that is long and complex in both scientific and industrial terms (Figure 3.2). The process is therefore vulnerable to interruption at a number of points. For example, raw material shortages or shortages in the production of refined and enriched reactor fuel would directly affect the capacity to produce power. Alternatively, interruptions might affect handling and exploiting of irradiated fuel, thereby increasing industry costs and generating political embarrassment from the accumulation of dangerous radioactive waste. In any event, it is clear that the capacity to design, build and install reactors is insufficient by itself to ensure the degree of independence and control which is the ultimate aim of French energy policy. For this reason, expansion of EDF's reactor programme has been closely matched by CEA's efforts to establish mastery of other stages in the nuclear fuel cycle (Carle, 1983). It is only in the context of both 'arms' of the industry that the extent of the French commitment can be appreciated.

FIGURE 3.3 NUCLEAR FUEL CYCLE OPERATIONS IN FRANCE

After EDF (1982e) Electricité d'orgine Nucléaire Aspects Techniques

Mining and milling

Although France is poor in conventional energy resources, she is well-endowed with naturally occurring uranium. She possesses 3.3 per cent of proven world reserves and 2.4 per cent of estimated potential resources. Determined exploration has extended the original mining areas in the Limousin throughout the Massif Central and into Hérault and Brittany (Figure 3.3). The operations are dominated by the CEA-owned Cogema Company, although small private companies, some of which are Cogema-controlled, contribute approximately twenty per cent of domestic output (CEA, 1983). The level of this output has been allowed to rise steadily from 2,000 tonnes in 1977 to about 3,000 tonnes at present, and there are plans to increase mining capacity to over 4,000 tonnes by 1990 (Zeigler, 1982; CEA, 1983). It is significant, however, that this increase does not match the increase in demand caused by the rapid expansion in reactor construction (Table 3.6). This reflects two interrelated features of uranium mining policy: first, a determination to conserve national supplies and second, consolidation of French control over foreign uranium resources. These features allow CEA to exploit comparative mining costs in other parts of the world and to avoid some of the growing domestic opposition to uranium mining which a more rapid expansion of the home industry would be sure to create (Schwarz, 1981: Ouest-France, 12th April, 1983).

TABLE 3.6 CONSUMPTION AND PRODUCTION OF URANIUM IN FRANCE (1980-1990)

Date	Nuclear Capacity (Mw.)	Uranium Needs (tonnes)	French Uranium Production Capacity (tonnes)
1980	10,000	3,400	2,630
1981	16,500	4,400	3,700
1982	22,000	5,100	3,900
1983	25,000	5,800	3,900
1984	33,000	7,300	3,900
1985	38,000	7,800	3,900
1990	60,000	9,000	4,050

Source: CEA (1983) (from "Réflexions sur l'approvisionnement en uranium de la France. Notes d'information du CEA", June, 1981.)

The principal areas of overseas investment are in the former colonies of Niger and Gabon where French influences, official and unofficial, remain very strong (Boyle and Robinson, 1981a). But in addition, French-owned interests have also been reported in Namibia and South Africa, in the USA and Canada, and in Australia, Indonesia and South America (inter alia: Lucas, 1979; Sweet, 1981; WISE, 1984). They are sufficiently extensive to allow the French to play an active role in the world uranium market and in the early 1970s they were largely responsible for creating a defensive cartel designed to challenge US domination of uranium supplies and to protect producer-interests (Pringle and Spigelman, 1983). In the long run, of course, it is possible to guarantee the political sympathies of those countries in which France has major investments. For the time being, however, these investments give the French valuable control over their own domestic resources and an important international voice in uranium exploitation.

Refining and enrichment

Uranium ore usually leaves the mines as 'yellowcake'. Before it can be fabricated into fuel it must be refined, and before it can be used in PWRs or weapons, the content of the fissile isotope U235 must be enriched. When the PWR programme was accelerated after 1973 uranium was refined at Malvesi, but enrichment facilities were restricted to a small military plant at Pierrelatte, south of Lyon, on the Rhône. The plant was operated by CEA (Figure 3.3; Lucas, 1979). Its limited capacity required overseas enrichment contracts to be arranged to meet the planned increase in demand from the expanded reactor programme. The most important of these contracts was signed with the USSR, partly out of considerations of cost but also out of the French distaste for US-dominated enrichment facilities in the West and the anti-proliferation conditions which the Americans imposed (Dumoulin, 1984). It was under the conditions of the Russian contract, which expires in 1990, that the Mont Louis was sailing with a cargo of uranium hexafluoride when she sank in the Channel in 1984 (see The Guardian, 31st August, 1984).

In the post-War period US domination of enrichment facilities presented a problem for all European nations with nuclear ambitions. Unfortunately, the investment required to construct a plant on a commercial scale is very high and even before 1973, talks had started which were directed to a co-operative European venture. The talks centred on two different technologies. Britain, West Germany and the Netherlands argued for a method which involved centrifuging uranium gas in order to spin off the heavier isotopes and to leave behind a residue which is enriched with the lighter fissile U235. Together with Italy, Spain and Belgium, on the other hand, the French preferred a gaseous diffusion process which works by passing uranium gas through a series of porous barriers. Because U235 is very light it passes through these barriers more easily than the heavier isotopes, thus allowing its collection at the end of the process. The two groups failed to reach agreement and in consequence two separate plants were built: the centrifuge at Almelo in Holland, and the diffusion plant at Tricastin on the Rhône.

The Tricastin plant, in which the French are the principal shareholders, is called Eurodif. It began enrichment operations in 1979 anticipating an annual output which would be sufficient to fuel a hundred 900 Mw. reactors. The reality, however, has differed from the ambition. A general downturn of nuclear programmes amongst Eurodif shareholders and contract holders, together with competition from the cheaper centrifuge plant at Almelo, means that it has functioned well below capacity, perhaps at only 60 per cent (Pringle and Spigelman, 1983; Samuel, 1983). Not only have plans to build a second enrichment plant been shelved, but the French are presently faced with the irony of honouring their Soviet contracts whilst simultaneously sustaining idle capacity at Eurodif. It may be, though, as Lucas (1979) has argued, that France is willing to sustain these economic penalties at her huge diffusion plant in exchange for the 'powerful market position' which her enrichment capacity ensures.

Enriched uranium is made into fuel elements at four separate sites: Romans, Annecy and Cadarache in the south-east, and Saclay near Paris. The fuel elements for the older generation of gas-cooled reactors and for the breeder reactor programme are manufactured by a Frency company called Société Industrielle de Combustibles Nucléaire (SICN). PWR fuel, however, is manufactured by a Franco-Belgian consortium, La France-Belge de Fabrication de Combustible (FBFC). Recently, a new Franco-Belgian company has been established, called Commox. It is jointly owned by Comega and Belgo-Nucléaire and its purpose is to develop a new fuel for light water reactors. The fuel, called Mox, consists of a mix of natural and spent uranium with between three and five per cent plutonium. Trials of the fuel have already started at Chooz and if they are successful a manufacturing plant will be built at Marcoule by 1992. One of the major attractions of this venture is that it will help to absorb some of France's growing stockpile of irradiated uranium and, in particular, it may help to obviate costly and dangerous storage of surplus plutonium (O'Dy, 1984b).

Reprocessing and waste management

The French have established a major industrial lead in reprocessing irradiated fuel from PWRs. British experience, though extensive, has so far been restricted to handling the 'cooler' fuels from gas-cooled reactors. The French operations are carried out by Cogema at Cap de la Hague in Normandy. Extensions to this plant, scheduled for completion by 1989, will provide a reprocessing capacity of 1,600 tonnes per annum which should certainly meet domestic requirements and provide a margin for treating spent fuel from overseas (O'Dy, 1983; CEA, 1983). The purpose of reprocessing is to separate unburned uranium and other potential fuel isotopes, like plutonium, from the useless by-products of the fission process: in theory this maximises the energy value of the fuel and minimises the residues of highly radioactive waste. Unfortunately, the technology is difficult and dangerous, especially for the treatment of the thermal oxide fuels used in light water reactors (Bunyard, 1981), and few countries have been willing to persist with its development. Even the USA has abandoned reprocessing in favour of long-term pool storage pending improvements in handling capabilities (Etemad, 1981). Moreover, although high level waste is reduced the spread of radioactive contamination during the process adds greatly to the total volume of intermediate and low-level waste.

In the face of the caution displayed elsewhere the enthusiasm of the French stands out in sharp contrast and invites two very different interpretations. First, it may be seen as an important indicator of their commitment to national control and independence in every sphere of the nuclear fuel cycle and, to some, it also proclaims the confidence of the industry in its own technology (Carle, 1983). In addition, it appears to reflect a conscious decision to secure commercial advantage by exploiting a major gap in the technologies of other nuclear states (Lucas, 1979; Pringle and Spigelman, 1983). This has been particularly important since overseas contracts have provided much of the capital needed for the La Hague developments. Countries like West Germany, Belgium, Holland, Sweden and Japan have been willing to buy off domestic responsibility and quieten opposition to nuclear expansion by exporting their reprocessing problems to France. It is clear that Britain is now determined to secure a share in this volatile export trade. There is no doubt that the thermal oxide reprocessing plant (THORP), currently under construction at Windscale, owes its existence in part to the determination of British Nuclear Fuels to capture some of the international market for reprocessing spent fuel rods from light water reactors (Bunyard, 1981). In addition, Franco-British co-operation now seems likely to lead to the development of a new facility at Dounreay designed specifically to reprocess fuel from European fast breeder reactors (The Guardian, 25th May, 1985; SCRAM, 1985; Stead, 1985).

In another view, however, the scale of the French commitment seems rash and precipitate. The concentration of international reprocessing in France and Britain threatens to make these countries the 'nuclear dustbins' of much of the Western world and it is quite certain that the operation of the La Hague plant, like the plant at Sellafield, has been far from satisfactory (Lucas, 1979; Bunyard, 1981). The uncertainties of the technology have been highlighted by a disturbingly high incidence of radiation leakage, accident and fire which have plagued the La Hague installations from their inceptions (*inter alia*: Lloyd, 1981; O'Dy, 1981; Tucker, 1981). Radioactive effluent discharge from the plant has led to protests from as far afield as the Channel Islands and the south coast of Britain, and labour relations with the industry's moderate trade union (Confédération Française Démocratique du Travail) have been severely strained by a parade of accidents involving radiation exposure (CFDT, 1980; The Guardian, 12th January, 1981; The Guardian, 4th February, 1981). Indeed, the union has emerged as a major critic of both plant design and organisation and of managerial competence. Unscheduled shutdowns, arising from a variety of causes, have prevented operation at full capacity and have increased costs which are anyway probably far higher than the industry is prepared to admit (Bunyard, 1981; Damveld, 1984; Table 3.7).

TABLE 3.7 FRENCH NUCLEAR FUEL COSTS (IN PERCENTAGES OF TOTAL COSTS)

Uranium mining and concentration	29%
Refining processes	2%
Enrichment	30%
Fuel fabrication	8%
Reprocessing and management of waste	31%
	100%

Source: CEA (1983)

Even if reprocessing is highly efficient a residue of treated materials remains extremely active and has no known use (O'Dy, 1983). Together with medium-and low-level waste these materials are the responsibility of the Agence Nationale pour la Gestion des Déchets Radioactifs (ANDRA), which is a rough equivalent of Britain's Nuclear Industries Waste Executive (NIREX). Since 1969, low-and medium-level waste has been stored in barrels on the surface or sunk in concrete lined trenches near La Hague. High-level waste is also stored there in liquid form in cooling ponds. It is estimated, however, that the capacity of the Centre de la Manche storage facility will be exhausted by 1990 and a sense of urgency now affects ANDRA's operations. A second storage site for high-level waste is located at Marcoule. Here the waste is vitrified before storage in air-conditioned warehouses, and similar facilities are now under construction at La Hague. The ultimate intention is to find geologically suitable deep storage sites for the output of these plants. (New Scientist, 23rd June, 1983; Marchand, 1984). Like its British equivalent, however, ANDRA has been embarrassed by the amount of opposition which has been mounted in the vicinity of proposed waste dumps (Blowers and Lowry, Chapter 6). Moreover, the handling of delicate issues seems sometimes to be clumsy and indifferent to local sensitivity: interest in a number of areas, in Haute-Garonne, les Bouches du Rhône and le Bassin de Lodève, for example, was first made public in a critical journal article in Science et Vie (reported in Le Parisien, 10th September, 1984). Not surprisingly potential host communities reacted with hostility to the lack of prior consultation or warning.

The difficulties of reprocessing and waste disposal were recognised by the Castaing Commission Report of 1982. In addition to a string of specific recommendations relating to operations and management at La Hague, the Commission also recommended a more cautious approach to handling spent fuel generally (Damveld, 1984). It was wary of the implications of long-term deep storage strategies for high-level waste, and was acutely aware of the political embarrassment which waste management failures were likely to generate. It recognised the disquiet which this area of operations aroused and recommended the establishment of a new research and development body charged with exploring waste disposal options, but freed from any affiliation with the existing vested interests of the French nuclear industry (O'Dy, 1983). To date, however, no action has been taken.

The economic critique

The reactor programme and the supporting fuel cycle operations represent a massive mobilisation of financial and industrial resources. It has been achieved through the concentration of power and the identity of interests between government, public utilities and French business. Moreover, it has found favour with both major political parties and its impetus has been effectively maintained through the Presidencies of Pompidou, D'Estaing, and now Mitterand (Boyle and Robinson, 1982). The arguments that have been deployed in defence of the programme have been pressed home vigorously: France does possess uranium but does not possess oil; the major oil-exporting countries are inherently unstable; there is massive export potential for the nuclear power industry; and, of course, nuclear energy is claimed to be cheaper than energy from fossil fuels (Mounfield, 1985). In the view of the establishment, to oppose nuclear development is to oppose the development of France. But, in fact, it is extremely difficult to calculate the real costs of the French nuclear expansion especially in view of the secrecy which pervades nuclear issues generally. EDF, like CEGB (Central Electricity Generating Board), insists that the industry is economically viable, even relatively cheap, and consequently of major benefit to the nation (EDF, 1983b). Its success in promoting this view and in campaigning for energy independence reflects a deeply-rooted insecurity about energy supplies that was gravely exacerbated in 1973. A number of recent studies, however, have presented alternative analyses of the nuclear investment programme which depict this spearhead of French technological achievement as an immediate burden and an impending liability of large proportions (Sweet, 1981; Lenoir and Orfeuil, 1983; Chemiot, 1984; Damveld, 1984). It begins to appear that the scale of investment has been excessive and that France faces an imminent, costly and embarrassing over-supply of generating capacity.

French policy has differed from British policy in one major aspect: it has gone beyond the concept of 'nuclear substitution', designed to diversify and stabilise energy-producing capacity, into a policy of 'nuclear electrification' under which the growth of electricity consumption has been deliberately encouraged. Between 1973 and 1981, for example, consumption rose by 51 per cent, which is double the increase for the EEC as a whole. At the same time, however, oil consumption declined little more in France (-21 per cent) than it did in the rest of the Common Market (-20 per cent). Official figures make much of the claim that nuclear electricity is about half the price of electricity from oil-fired plants. Unfortunately, they rarely point out that electricity from any source is very much more expensive than coal, gas or oil used directly. In 1983, for example, Lenoir and Orfeuil estimated that prices per Kwh, to industry were 12.62 centimes for coal, 14.3 to 15.3 centimes for gas, 17.3 to 18.5 centimes for oil, and 29.1 centimes for electricity. The lack of industrial demand for electricity, therefore, is not surprising despite EDF's promotional expenditure of 400 millions francs in 1983, and a likely one billion francs in 1984 (Lenoir and Orfeuil, 1983).

Compensation for stagnating industrial sales has been found in a more compliant and captive domestic market. France, for example, is the only country in Europe which still actively encourages electric space-heating, and there is now little alternative for new homes. Together with an expansion in sales of electrical home appliances, this has led to an increase in domestic power consumption of over 130 per cent since 1973. Once again, however, the consequences are neither direct nor simple. Reactors are designed to run at a constant output level: they cannot easily be adapted to meet the peaks which domestic demand entails. These are still largely met from conventional sources, especially from coal. Indeed it is largely in consequence of this inflexibility that the Government has stabilised domestic coal output, encouraged the expansion of coal imports, and authorised the construction of new coal-fired power stations and the conversion of stations previously fired by oil (Etemad, 1981; News from France, 1980). As a result barely 10 per cent of electric heating is met from nuclear output, and coal, oil and gas, which could be used cheaply and directly, are instead converted into expensive secondary power. Because of EDF's commercial policies, designed to sell enough electricity to justify their expensive construction programme, household electricity bills rose by up to 40 per cent in 1983 (Lenoir and Orfeuil, 1983; Chemiot, 1984).

Demand forecasting errors, based on the now unrealistic expectation that electricity consumption will double every ten years, have already created a significant excess of installed generating capacity which will continue to increase as new plants begin operation (Damveld, 1984; SCRAM, 1984). EDF is now forced to assume that its nuclear plants will be under used, operating for only 4,000 hours per year (out of a possible 8,760) in 1990. This means reduced efficiency and increased electricity costs. Various strategies are being explored to cope with this over-capacity. The EDF campaign to stimulate industrial use of electricity is one, and premature shutdown of approximately 2,800 Mw. (26 units) of oil-fired power stations is another. Moreover, France exported 13.5 Kwh. of electricity in 1983 and hopes to increase this to 20 billion Kwh. annually (WISE, 1984). There are contracts signed or under negotiation with the Netherlands, the Channel Islands, West Germany, Switzerland, Spain and Italy (Damveld, 1984). In addition, CEGB now buys off-peak French electricity via a new cross-Channel link (The Guardian, 8th February, 1984; Gibb, 1985). Because of differences in peak loading times the new cable connection offers both countries limited flexibility in the exploitation of power producing capacity and marginal generating costs (Gibb, 1985).

Meanwhile EDF made losses of 4.4 billion francs in 1981, 7.9 billion francs in 1982 and 6 billion francs in 1983 (the 1982-3 reduction is a result of tariff increases and exports). Lenoir and Orfeuil (1983) accuse the State of playing accounting games to disguise the scale and seriousness of EDF's financial situation. Moreover, EDF debts were expected to reach 210 billion francs by the end of 1984. These have increased from 100 billion in 1980. Nuclear investments account for more than half of the borrowing and EDF recently borrowed $300 million on the international bond market to refinance its 70 billion francs external debt. This foreign debt has become increasingly expensive because of the weakness of the franc and because so much of it is in US dollars. The scale of this commitment means that EDF is now the most important borrower on the US money market and the third in international money markets. Far from solving France's balance of payment problems, therefore, the huge scale of investment in the nuclear programme has made them significantly worse. The financial implications are the same whether it is energy or currency which is purchased abroad (Lenoir and Orfeuil, 1983; Damveld, 1984).

Unease and opposition

It is in the nature of all nuclear developments that they attract opposition from some quarter. Given the spread and the scale of the French programme it is not surprising that objections have been made on a variety of fronts, for a variety of reasons. The questions raised by the nuclear commitment penetrate the fabric of French life at almost every level, raising questions of morality and principle, of political economy, and of practical everyday living. Yet these questions have had little real impact on the activities of EDF and CEA who have employed markedly different tactics to ensure the integrity and continuity of their energy plans. The tactics fall into three broad, but not mutually exclusive, categories. First, at a local scale, they have relied upon public relations skills and financial inducements to override the vague uncertainties of host populations. Second, attempts in other quarters to question the impact of the nuclear investment on the French economy, on other sectors of industry, and on domestic consumers have been met with secrecy, or indifference, or blunt counter-assertion. Finally, the anti-nuclear movement, which challenges the proliferation of nuclear technology and its military connections from a moral standpoint, has been bludgeoned into disarray by the crude exercise of State authority. We have already outlined something of the economic critique; here we turn to a short consideration of the relationship of the industry with host communities and with the anti-nuclear movement.

Local opposition and incentives

The reactor programme's rapid expansion has faced EDF with the problem of selecting many sites in a relatively short space of time. This is a potentially contentious and always problematical issue, since it involves a mix of decisions on factors which are physical and technical on the one hand, and economic or socio-political on the other. Striking an appropriate balance has been a major difficulty but one which appears in retrospect to have been accomplished with considerable skill, quietening or placating opposition without seriously compromising basic siting strategy.

The crux of this strategy is simple: it is to minimise transmission costs and right-of-way negotiations for power lines by locating reactors as close as possible to centres of demand. In most cases this requires inland riverine locations rather than coastal ones. It also carries a concomitant need to employ 'closed-loop' cooling via cooling towers instead of the less obtrusive 'once-through' system that coastal sites and the occasional larger river permit (EDG, 1982e). Inevitably this has inspired a reaction against the impact of cooling towers in otherwise pleasing rural environments, partly on aesthetic grounds and partly out of a perceived threat to local tourist industries. In this context EDF appears extremely flexible: it has been prepared to compromise by way of design adaptations calculated to reduce the visual impact of its developments. At St. Laurent des Eaux, for example, cooling towers have been scaled down from the standard 165 metres to 120 metres by installing pumps to reinforce natural cooling draughts. More extravagantly, at Chinon in the wine-growing area of the Loire, the 'towers' have been reduced to only 29 metres by relying entirely on pump-generated cooling (EDF, 1982a; Carle, 1983). At Paluel, the entire reactor complex has been masked from view by sinking it into a cutting in the chalk cliffs. Cosmetic adjustments of this kind, although costly, have enabled EDF to present a public image of an institution which is both sensitive and environmentally aware. It is an image which it fosters assiduously.

The general public also have less well-defined apprehensions, particularly those who live near proposed reactor sites. There is an unease which stems inevitably from the more macabre dimensions of nuclear technology: the risk of accidental leakage, for example, and the insiduous consequences of radioactive emissions. It takes the form of imprecise doubt rather than an articulate opposition, and it has been approached by EDF with a skilful blend of public relations and tangible inducements. By their practicality and immediacy these inducements are designed to outweigh the vague disquiet of host populations. To begin with, EDF is a major employer. In a period of recession and growing unemployment the nuclear industry provides over 150,000 jobs and it has the general sympathies of the trade union movement. Locally, it channels a large proportion of construction work into small firms with all the employment and economic benefits that this entails. At Nogent sur Seine, for example, 38 per cent of all site contracts have been given to local firms, and at Chinon 85 per cent of civil engineering labour is locally based (EDF, 1983a). In addition to these attractions and to the stimulation of existing business and the encouragement of new business, EDF also reduces electricity tariffs for host communities. These reductions have been guaranteed until at least 1990. EDF also pays a direct tax (*the taxe professionnelle*) and a rating levy (*the taxe foncière*) to the communes in which the installations are built and to the département as a whole. The scale of these payments can be considerable. On the 4,000 Mw. installed at Bugey, for instance, the combined payments for 1982 amounted to the equivalent of around £6 million (EDF, 1983a).

EDF presents these inducements in a public relations context which depicts a benevolent employer and an enlightened partner in the local community (EDF, 1983a). It stresses the benefits to be obtained from reconstructing schools, enlarging sporting facilities, and improvements in the quality of life which are apparent to the most casual onlooker. A less charitable view, of course, might see in such a policy little more than the crude buying-off of long-term uncertainties with a seductive coinage of more immediate but more superficial value than a nuclear-free countryside.

The anti-nuclear movement

The anti-nuclear movement in France, as elsewhere, exists as a fragile congeries of disparate elements who share a common opposition to nuclear developments of any kind. At a risk of over-simplification their objections are best described as 'moral' and 'principled', rather than 'technical' and 'rational', and their modes of argument are humanist and emotional rather than scientific and analytical. They embrace a variety of political philosophies and they are drawn from all walks of life and driven by motives which are often internally conflicting. Such unity as the movement is able to achieve finds its most concrete expression in physical demonstrations designed to display implacable opposition to a policy which is seen to be against the best interests of French society and of people in general. Far from being a source of strength, however, this inherent diversity is a cause of weakness and the internal tensions to which it gives rise prevent the establishment of a coherent, sustained, and politically viable alternative to national energy policy (Touraine *et al*., 1980). It is true that in the case of Plogoff in Brittany a large reactor development was abanoned in the face of bitter opposition. But in a real sense that was the outcome of a political concession made to Breton identity by President Mitterand during his election campaign in 1981 (Boyle and Robinson, 1981b). Elsewhere, the anti-nuclear movement has made little impression on an equally implacable and much more effectively organised nuclear industry.

Not least of the difficulties which plague the anti-nuclear movement lies in conflicting attitudes towards the use of violence. Pacifists of all political hues are ranged alongside minority groupings of the extreme left and descendants of the French anarchist tradition, some of whom advocate, or at least acquiesce to, acts of violent physical confrontation and even sabotage as being the only effective ways of registering the strength of their opposition (Lucas, 1979). For these groups, of course, this reflects moral and political disenchantment with the French social order and the necessary espousal of revolutionary reform. But for many others, the occasional outbreaks of violence which have been a disturbing feature of anti-nuclear protest, have much more mundane origins. They are an expression of frustration at the intransigence of French bureaucracy, the arrogance of its institutions, and the lack of a suitable forum in which matters of principle and matters of immediate local concern can be effectively voiced. In particular, they reflect the inadequacy of the public inquiry procedure (l'enquête d'utilité publique) which is designed to settle compensation claims rather than to allow the arbitration of deeply divided viewpoints (SOFEDIR, 1978; Le Diouron et al., 1980; Armstrong, Chapter 4 this volume). Moreover, as several studies have indicated, the behaviour of the police - invariably the Compagnie Républicaine de Sécurité (CRS) - has been marked by excessive brutality towards the legitimate demonstration of dissent (Lucas, 1979; Touraine et al., 1980; Boyle and Robinson, 1981b).

Our own conversations with members of the anti-nuclear movement suggest that since the demonstrations against the commercial fast reactor at Creys-Malville in 1977, when one of the demonstrators was killed and scores of others were injured, the united front that was evident then has become fragmented and fallen into disarray. The movement now sees opposition to the PWR programme as a lost cause and it is quite certain that the CRS is much feared. Apart from the limited and occasional successes of the Ecology Party under Brice Lalonde, political assent to the nuclear industry's ambitions has been readily won and easily maintained (McDonald, 1980; Le Figaro, 8th October, 1981; Smyth, 1982). Even the public debate invited by Mitterand in 1981 raised scarcely a ripple and, despite the scaling down of the programme in response to revised energy forecasts, the essential policy has emerged unscathed (Le Figaro, 26th November, 1981; The Times, 27th November, 1981; Boyle and Robinson, 1982). Confident in its unity with the State, in backing from all the major political parties, and in the acquiescence of the general population, the nuclear industry pays scant regard to an incoherent and demoralised anti-nuclear lobby. We think that, with the exception of occasional protests arising out of particular incidents, the anti-nuclear movement in France has been weakened beyond recovery. National and international organisations, such as Les Amies de la Terre and the Groupement des Scientifiques Concernés, are no longer regarded as a serious threat to the impetus of the PWR programme. The fast reactor, reprocessing and waste management will continue to arouse disquiet but, in principle, the battle for 'tout nucléaire' has been won. Only its consequences have yet to be revealed (Masse Critique, 1983, 1984; Assises Européennes Contre la Surgénération, 1984; Campagne Pour l'Arrêt de Malville, 1984).

The French lesson

We have emphasised two features of the French nuclear programme which cause it to be distinctive. The first is its scale and the speed with which it has been tackled. The second is its completeness in the sense that it embraces every aspect of the fuel cycle. It is an undeniably impressive achievement, at least in material terms, and it may be paraded as a courageous and visionary exploitation of a new technology to solve an old problem. But as Lucas (1979) has shown, the level of commitment which has been required is substantially an expression of the 'special character' of French culture and institutions. In this sense, its scale articulates something of the self-image of a particularly proud nation and of that nation's leading figures, whilst the simultaneous vigour of its prosecution exposes the extreme concentration of power that is the hallmark of French bureaucracy.

The British nuclear industry would like to follow the French lead, at least in developing technological capabilities commercially. In many respects, however, Britain appears only now to be reaching the stage that was passed in France fifteen years ago. The efforts of the Central Electricity Generating Board to secure authorisation for Westinghouse PWRs come at a time when the French are abandoning them in favour of domestically-designed reactors. Disagreements within the British industry on the merit of different reactor types echo the much earlier disputes between CEA and EDF. The shift to diversification, with light water reactors supplementing indigenous gas-cooled types, contrasts with standardised French reactor systems, which have stabilised and strengthened their construction industry and enlarged its potential export capacity. It seems unlikely that the British can ever match the French in commercial reactor sales except as middlemen for American technology. Indeed it may be that only in the environmental minefield of reprocessing is there a substantial share in the commercial spoils of international reactor proliferation.

The lack of progress in the British industry reflects a number of fundamental differences between the two countries. One is relatively energy-rich while the other is energy-poor, and the insecurity experienced by the French has been much less marked on the other side of the Channel. On the contrary, the British have enjoyed a flexibility in energy options that is enviable. Sectors of the nuclear industry that are presently State-controlled are even being prepared for privatisation in sharp contrast to the purposeful extension of the EDF/CEA alliance into company domination (The Economist, 7th July, 1984; Guihannec and Rebattet, 1984). But beyond commercial questions, and beyond questions of energy supply and demand, lie differences in institutional power and control and differences in political and planning processes that expose the actions of the British industry to far more democratic scrutiny. The anti-nuclear lobby, though less volatile than its French equivalent, is stronger and more consistent. Faced with the luxury of choice the decision-making process slows to the point of meandering and energy planning wallows in incoherence. (See Armstrong, Fernie and Openshaw, Blowers and Lowry, this volume Chapters 4, 5 and 6.) This has two, sometimes, conflicting, consequences. Precipitate action on a major scale is checked, but the price of these checks may be slackening technical impetus and loss of commercial advantage.

It is too early to attempt a definitive perspective on the French nuclear industry and on its full significance in international terms. But the great design which emerged during the 1970s is beginning to look conspicuously flawed. Even the Rapport Joséphe of 1983 recorded concern at the over-capacity of the nuclear industry and the fear that in the 1990s the problem for the French will lie not in securing energy, but in securing the monolithic structure of an industry created in their ambitious search for independence (Bonazza, 1983; New Scientist, 28th July, 1983). For Britain, much more richly endowed with conventional energy, the emerging French condition should urge caution. So demanding, so uncertain, and so expensive a technology is not immediately necessary on a large scale: prudence favours the tortoise.

Acknowledgements

We would like to acknowledge the friendliness and assistance given to us by individual EDF employees and by representatives of anti-nuclear movements in Lyon. We would also like to thank Jean Mellor for typing, Graham Bowdon for drawing the maps, and the University of Manchester for financial assistance with fieldwork in France. Finally we want to thank David Pepper and Andrew Blowers for their constructive comments on earlier drafts of the chapter.

REFERENCES

ARDAGH, J. (1982) France in the 1980s, Harmondsworth, Middlesex: Penguin.

Assises Europeennes Contre La Surgeneration (1984) Compte Rendu des Assises Europeennes Contre La Surgeneration, 26-27th May, 1984, Lyon.

BONAZZA, P. (1983) 'Energie: le dosage difficile', L'Express, 22nd July, 1983, 42-44.

BOYLE, M. J. and ROBINSON, M. E. (1981a) 'French nuclear energy policy', Geography, 66, 4, 300-303.

BOYLE, M. J. and ROBINSON, M. E. (1981b) 'Plogoff says no to nuclear power', Geographical Magazine 53, 11, 681-3.

BOYLE, M. J. and ROBINSON, M. E. (1982) 'A further note on French nuclear energy', Geography 67, 2, 148-9.

BUNYARD, P. (1981) Nuclear Britain, London: New English Library.

Campagne Pour L'Arret de Malville (1984) L'Aberration, 10th May, 1984.

CARLE, R. (1983) 'How France went nuclear', New Scientist, 13th January, 1983, 84-6.

CEA (Commissariat a l'Energie Atomique) (1983) Quelques Informations Utiles, Paris: Departement de Relations Publiques et de la Communication.

CFDT (Confederation Francaise Democratique du Travail (1980) Le Dossier Electro-Nucléaire, Paris: Seuil.

CHEMIOT, D. (1984) 'Le nucléaire pervers', Alternatives Economiques, March, 22-3.

DAMVELD, H. (1984) Electricity Costs in France, Amsterdam: WISE.

DUMOULIN, J. (1984) 'Nucléaire: l'heritage des années folles', L'Express, 7th September, 1984, 42.
EDF (Electricité de France) (1982a) Le Programme Electro-Nucléaire Française, Paris: EDF.
EDF (1982b) The French Nuclear Electricity Programme, Paris: EDF Construction Division.
EDF (1982c) Energies et Environnement, Paris: EDF.
EDF (1982d) Electricité d'Origine Nucléaire Aspects Sûreté, Paris: Service de l'Information et de Relations Publiques.
EDF (1983a) Les Consequences de l'Implantation d'une Centrale Nucléaire sur l'Economie Régionale et l'Emploi, Paris: EDF Direction de l'Equipement.
EDF (1983b) Les Centrales Nucléaires, Paris: EDF Service de l'Information et de Relations Publiques.
ETEMAD, S. (1981) 'Handle with care', The Guardian, 22nd January, 1981.
GALLACHER, J. (1984) 'Le plutonium nouveau est arrivé', SCRAM Energy Journal 43, August/September, 1984, 10.
GIBB, R. (1985) 'Helping to turn each other on', Town and Country Planning 54, 1, 16-17.
GUARDIAN, 12th January, 1981, 4th February, 1981.
GUARDIAN, 31st August, 1984, 25th May, 1985.
GUIHANNEC, Y. and REBATTET, A. (1984) 'Creusot-Loire et le mal Française', L'Express, 13th July, 1984, 41-4.
LE DIOURON, T., CABON, A., de LIGNIÉRE, C., PERAZZI, J. C., THEFAINE, J. and YONNET, D. (1980), Plogoff: la revolte, Le Guilvinec: Editions Signor.
LENOIR, Y. and ORFEUIL, J. P. (1983) '10 ans de programmes nucléaire: EDF devient un fardeau pour la France', Science et Vie 194, 22-23 and 164-170.
LLOYD, A. (1981) 'Fire at French nuclear fuel plant leaks radiation', New Scientist, 15th January, 1981.
LUCAS, N. (1979) Energy in France: Planning Politics and Policy, London: Europa.
MARCHAND, J. (1984) 'Nucléaire: le gouffre aux déchets', L'Express, 14th January, 1984, 65-66.
MASSE CRITIQUE (1983 and 1984 various editions) Mensuel de la Coordination Nationale Antinucléaire: Lyon.
McDONALD, J. (1980) 'Environmental concerns and local political initiatives in France', Geo-

graphical Review 70, 3, 343-52.
MOUNFIELD, P. R. (1985) 'Nuclear power in Western Europe: geographical patterns and policy problems', Geography 70, 4, 315-327.
NEWS FROM FRANCE (1980) 7, 2, 6-7, August.
O'DY, S. (1983) 'Nucléaire: que faire des déchets? 'L'Express, 14th January, 1983, 59-60.
O'DY, S. (1984a) 'Nucléaire: le marriage de raison', L'Express, 2nd March, 1984, 59-60.
O'DY, S. (1984b) 'Du plutonium dans le combustible', L'Express, 23rd November, 1984, 67.
PRINGLE, P. and SPIGELMAN, J. (1983) The Nuclear Barons, London: Sphere Books.
SAMUEL, P. (1982) Facts about the French PWR Programme, Paris: Les Amis de la Terre.
SAMUEL, P. (1983) Nuclear Power in France: A Historical Account, Paris: Les Amis de la Terre.
SCHWARZ, W. (1981) 'The anti-nukes plan to break the uranium network', The Guardian, 14th March, 1981.
SCRAM (Scottish Campaign to Resist the Atomic Menace) (1984) SCRAM Energy Journal 43, August/September, 1984, 5.
SCRAM (1985) SCRAM Energy Journal 49, Supplement on Dounreay, August/September, 1985.
SMYTH, R. (1982) 'Mitterand turns off greenpower', The Observer, 24th January, 1982.
SOFEDIR (Société Française d'Editions et d'Information Régionales) (1978) L'Energie Nucléaire: Le Project de la Centrale Nucléaire de Nogent Sur Seine Palaiseau: SOFEDIR.
STEAD, J. (1985) 'Nuclear dustbin project ends 35 years of peace and trust', The Guardian, 10th June, 1985.
SWEET, C. (1981) A Study of Nuclear Power in France, Energy Paper No. 2, Department of Social Sciences, Polytechnic of the South Bank, London.
TOURAINE, A., HEGEDUS, Z., DUBET, F. and WIEVIORKA, M. (1980) La Prophétie Anti-Nucléaire, Paris: Seuil.
TUCKER, A. (1981) 'Liberty, equality and radiation?' The Guardian, 12th February, 1981.
WISE (World Information Service on Energy) (1984) Country Report, Paris: WISE.
ZEIGLER, V. (1982) 'Les resources en uranium', Revue de L'Energie 347, 831-41.

Chapter Four

DEMOCRACY AND THE SIZEWELL INQUIRY

Jennifer Armstrong

Introduction

The Sizewell Inquiry into the Central Electricity Generating Board's (CEGB) proposal to build a Pressurised Water Reactor (PWR) at Sizewell on the Suffolk coast ran from 11th January, 1983, to 7th March, 1985, a total of 340 working days. It outran by almost 100 days the lengthy Third London Airport Inquiry. Precise costs are not available but the CEGB which, unusually for the proponents at an inquiry of this sort, footed most of the bills, is thought to have spent some £20 -25m. More was at stake, however, than time and money. Leaving aside the implications for future investment in nuclear power generation, the Sizewell Inquiry, focusing on one of the most controversial of current issues, put to the test the Government's commitment to public involvement in the decision-making process. It also took place in the wake of a serious accident at another PWR (Three Mile Island) in the United States of America which shook public confidence in nuclear power in general and in this type of reactor in particular (see Chapters 5 and 12 in this volume). During the course of the inquiry, two possible sites were named for the long-term storage of nuclear waste, at Billingham in Cleveland and at Elstow in Bedfordshire. These announcements gave rise to immediate, all-party opposition (see this volume, Chapter 6). During the second half of the inquiry, a national coal miners' strike took place, occasioned by the announcements of pit closures. Inevitably, this added force to the proponents' arguments that nuclear power, despite its economic, safety and environmental uncertainties, was a vital component in any plan to reduce dependence on coal.

Democracy and the Sizewell Inquiry

This chapter briefly identifies the Government's publicised intentions for the Sizewell Inquiry, placing them alongside the diverse aspirations of its participants. It looks at how the two match up, and how the inquiry system fared in meeting the demands made upon it. The experience of the Sizewell Inquiry clearly points to the need to reform the procedures by which major planning proposals are subjected to public scrutiny.

The Government's Conflicting Objectives

Opposition to nuclear power, which was almost nonexistent when the first and second programmes of nuclear power stations were agreed in the 1950s and 1960s, grew rapidly in the 1970s, finding its first major focus in the 1971 proposal to build a spent fuel reprocessing plant at Windscale in Cumbria. In the aftermath of that contentious debate and the Three Mile Island accident, the newly elected Conservative Government was only too well aware that its 1979 announcement of a large nuclear power programme based on the PWR, was heading for trouble. First and foremost, it wanted to increase substantially the contribution which nuclear power made to total electricity generation. But it also wanted to be seen to be taking account of a growing public concern. These two potentially conflicting objectives can be identified in the statement made to the House of Commons in December 1979, by the Secretary of State for Energy, concerning 'our nuclear programme'. The House was told that 'safe nuclear power and strong nuclear industry are essential to this country's energy policy ... and we have made clear our wish that subject to the necessary consents and safety clearances, the PWR should be the next nuclear power station order, with the aim of starting construction in 1982 ... The future success of our nuclear programme is of great importance to the prosperity and security of this country. I ask all concerned to give their active support to the <u>decisions</u> which I have announced' (my emphasis).

No one reading the Secretary of State's speech could be unsure of the strength of the commitment to nuclear power in general and, subject to licensing, the PWR in particular. Furthermore, both the scale of the programme announced ('at least one new nuclear power station a year in the decade from 1982'), and the absence of any mention of how it might relate to other means of generating electricity strongly suggested that nuclear power had become synonymous with energy policy. No connections were made with the 1974 Plan For Coal, the then current feasibility studies on the Severn tidal barrage or, indeed, the British-designed nuclear plant, the Advanced Gas-Cooled Reactor (AGR).

As for the public inquiry into the first PWR, which was announced in advance of any site being named, one might conclude from the above statement that the context of the proposal was well established and that only matters of detail would be left for public discussion. The Secretary of State, however, went on to speak of the need for 'the fullest explanations and discussions to inform (the Inquiry). I am assured that all the principal safety documentation relevant to the initial licensing will be made available to the Inquiry and published'.

Turning elsewhere, we see a different picture. Just before the above Parliamentary announcement, Minutes of a Cabinet sub-committee meeting held in October 1979, were leaked to the press. These record a desire to adopt a 'low profile approach' to the nuclear programme since 'opposition to nuclear power might well provide a focus for protest groups over the next decade'. In discussing the tactics for the first inquiry, it is noted that 'there was a danger that a broad ranging inquiry would arouse prolonged technical debate between representatives of different facets of scientific opinion' - the assumption clearly being that such debate would be unwise.

Two issues - nuclear power in the context of national energy policy - and, more specifically reactor safety - illustrate the problems and tensions which arose when the nature and scope of debate sought by some factions of the public at the Sizewell Inquiry - and indeed promised by the Government - did not in fact materialise.

Energy Policy

In October 1980, the Sizewell site was named and in July 1981 the Secretary of State for Energy announced issues which he would take into account 'in considering whether the proposed power station should proceed'. These included: 'The CEGB's requirement for the power station in terms of the need for secure and economic electricity supply and having regard to the Government's long-term energy policy' (my emphasis). Potential inquiry participants welcomed this statement, believing it would allow a proper testing of the 1979 assertion that 'nuclear power is essential to this Government's energy policy, even though the inquiry, strictly speaking, was about local planning issues'.

By 1980 that already uncertain policy background was complicated by the downward revision of electricity demand projections, and some serious criticism of the CEGB's planning and investment strategies.

One source of such criticism was the 1981 report of the House of Commons Select Committee on Energy which had taken place as its first subject of enquiry the Government's 1979 announcement on nuclear power. One of its main conclusions was that, while investment in energy conservation may be as cost effective as investment in new generating plant, no comparative work had been done. The Committee also suggested investigating the conversion of oil-fired plant to dual firing, better integration with the South of Scotland Electricity Board (SSEB) which has a very large electricity surplus, the potential savings due to privately generated electricity, and a reduction in the amount of 'spare' electricity available on the grid to meet contingencies. Thus, they were critical of the apparent basis of the scale of the nuclear proposals, recommending that the case for each new station be evaluated on its merits 'and not as part of a pre-determined programme'.

In the same year, the Monopoly and Mergers Commission reported on the CEGB, reaching similar conclusions, accusing the Board of acting against the public interest by promoting nuclear power without reference to actual electricity demand.

It is not hard to see, therefore, why objecting parties wished to challenge matters of energy policy, and the Department of Energy's announcement at the first Sizewell pre-inquiry meeting that, contrary to normal practice, its witnesses would be open to questioning on the merits of policy was welcomed. In the event, eight days' cross-examination of that witness did little to clarify matters (Inquiry, days 40-47). His role, as he saw it, was only to 'sketch the backcloth of the Government's general approach to energy policy'. A free market situation was favoured which prevented any 'set blueprint of energy development' although 'diversity and security of supply' which included 'an appropriate nuclear component' was thought to be the key. But he did not expand upon the thinking behind the clear cut and ambitious nuclear programme which had given rise to the Sizewell Inquiry, and there was no discussion of its details.

The fact that the inquiry was essentially a local planning inquiry, held in response to a particular planning application, greatly helped the Government and the CEGB in their efforts to restrict questioning to this one-off proposal. Yet between the 1979 announcement and the start of the inquiry several follow-on sites to Sizewell had been named. These were Hinkley Point in Somerset (a third station here), Druridge, Winfrith Heath in Dorset, and Dungeness in Kent. A connection between these sites and the Sizewell Inquiry is identified in the already cited 1981 report of the Select Committee on Energy, one recommendation of which is that 'future public inquiries should be site-specific and not re-open the wider issues of principle covered at this first inquiry'. This recommendation is noted in the Government's 1982 White Paper on nuclear power.

One of the sites named is Druridge Bay in Northumberland where, at the time of the Sizewell pre-inquiry meetings, the CEGB was opening negotiations to purchase the land. Nothumberland County Council's view was that if the site was ever the subject of a public inquiry, the Council would wish to raise issues of a non-site-specific nature. After lengthy and inconclusive correspondence with the Department of Energy, the County Council's solicitor sought guidance from the Inspector of the Sizewell Inquiry (pre-inquiry meeting, June 1982, p.31 of transcript), explaining that 'the CEGB proposes to build a nuclear power station, whereas my Authority feels that the construction of a coal-fired station would be more appropriate'. Was this, he asked, a matter which his Council would be able to raise if the Druridge Bay site were the subject of an inquiry, or would it be regarded as one of the wider issues of principle not relevant to a site-specific inquiry? He expressed the hope that such matters would in fact be able to be raised at future inquiries so that time and money would not need to be spent in attending the Sizewell Inquiry, some 300 miles from Northumberland.

Neither the Inspector nor the Department of Energy was able to give the Council the reassurances it sought. The Inspector could not comment on a future inquiry for which he might well have no responsibility. The Department of Energy's solution was to assert that the scope of future inquiries lay entirely with future inspectors. Deciding it had no choice, Northumberland County Council submitted a proof of evidence to the Sizewell Inquiry in which it criticised the Department of Energy for failing to explain the Government's view on the future of the coal industry, on the role of coal in future power station fuelling and on the effects of a nuclear power programme on the coal industry. The Department's answer (day 40, p.64) was to argue firmly that 'it is the case that the Government has no such programme' and 'has not formulated an intention beyond Sizewell'. In other words, the 1979 statement relating to a series of up to ten reactors was no longer a 'policy'.

Wansbeck District Council in Northumberland - another party to the Inquiry drawn in by the Druridge Bay site - picked up these anomalies in its closing speech, concluding that 'the Council is of the opinion, despite contrary statements by both the Secretary of State and the Board, that this Inquiry is as much into a programme of PWRs as it is into a single station' (day 319, p.108). It pointed to the 1982 White Paper's view that 'it will be difficult for the industries which supply nuclear power stations to keep their costs down unless they have reasonable prospects of future orders ...' and argued that 'what has changed since 1979 is not the underlying intention (to build a number of PWRs) but the way in which it is presented to the public ... the denial of the existence of a programme of nuclear power stations does not help this inquiry'.

These conclusions were supported by the fact that, in the month before the Inquiry ended, the CEGB's main policy witness gave interviews to the Press (e.g. Guardian, Financial Times, 25/2/85) in which he stated that the Board was not interested in building Sizewell without the understanding that further PWRs - possibly four or five - would follow in fairly quick succession.

The Reactor Safety Case

A second example of the gap between pre-inquiry promises and what actually happened at the Inquiry concerns the issue of reactor safety. 'I am assured' the Secretary of State for Energy had said in his 1979 statement, 'that all the principal safety documentation relevant to the initial licensing will be made available to the Inquiry and published'. This and other assurances, including that given by the Nuclear Installations Inspectorate (NII) itself, led participants to believe that the licensing process and the public inquiry would take place sequentially and not concurrently, with a reasonable period between the two (the Select Committee on Energy recommended at least 4 months) in order to allow a proper assessment of the complex safety material.

As early as 1981, however, doubts began to be voiced over whether such a timetable could be adhered to. By November of that year the NII said that 'there is no way in which we can have completed our full studies' in time for a public inquiry starting in October 1982. In January 1982, Friends of the Earth (FOE) wrote to the Secretary of State for Energy, reminding him of his promise to publish the full safety report and expressing concern over the rumoured publication of only a 'status report' which would not give detailed consideration to all the major safety issues. When a January 1983 start for the Inquiry was announced it was clear that the NII would not be in a position to license a PWR by then, let alone 4 months before.

FOE, which led the opposition on safety grounds, twice applied (on days 13 and 66) for an adjournment of the Inquiry because full safety information was not available. The CEGB's response was to stress that the inquiry was being held to deal with the planning application and not the licensing of the reactor: the latter was described as an 'iterative process', always open to change and difficult to present to the public at any particular moment in time. In replying to the adjournment requests (on days 21 and 76), the Inspector was of course well aware of the Government's promises that the reactor would have reached a licensable design stage before the Inquiry began. He acknowledged that there was 'considerable force' in FOE's request and that 'they had been led to believe that they would be able to rely on a much greater, fuller and earlier presentation of safety evidence than has been the case'. He concluded, however, that this did not provide him with the grounds to adjourn the Inquiry. Like FOE, he too was handicapped by the lack of progress on the reactor licence which, even when the Inquiry ended, was not expected to be issued until the following year.

Financial problems of objectors

The Government promised not only a full enquiry but also a fair one. A measure of the latter might be equality of opportunity to participate in the proceedings. At a major public inquiry effective participation can cost a great deal. Thus, the absence of funding for objecting parties was widely regarded as unjust: not only did the Government show its support for the proposal before the Inquiry began, but any material challenges could be quashed by a denial of financial aid. Some potential objectors such as the environmental body Greenpeace, withdrew from the arena before the Inquiry started, denouncing it as an elaborate piece of window-dressing. For those more committed to the established decision-making processes, the choice was less simple - although many were acutely aware of the problems of poverty at a major inquiry after the marathon Windscale debate.

At Sizewell's pre-inquiry meetings they argued that without funding they could not present proper cases and that the purpose of the Inquiry could not therefore be achieved. But in response to a 6-page letter from the Inspector, setting out objectors' arguments and asking for the provision of funding to be reconsidered, the Secretary of State (in a letter issued by a Department Energy Press Officer, 23/9/82) remained 'unconvinced' that a lack of financial assistance 'will make it impossible for the Inquiry to be full, fair and thorough'. He went on to say that 'not only objectors but also the taxpayer has an interest in the conduct of public inquiries. The Government must have regard to both interests'. Another line of defence was that funding of objectors would constitute a duplication of the NII's role, a body which was 'uniquely equipped ... to make a thorough and independent examination of the safety issues' (but which was not in a position to do so at the time). Finally he considered that 'the statutory provisions governing the Inquiry are themselves designed to safeguard the general public interest, including the interests of objectors'.

The response revealed a great deal about the Government's perception of the content and conduct of public inquiries and of the nature of the opposition, and begged many questions. The NII, however well-equipped (and it was not well-equipped on the PWR work) was actually involved with only a small area of objectors' total concerns. Neither can the setting up of an inquiry be regarded as anything more than the provision of a mechanism for public participation: the argument that its mere existence could somehow protect the public interest was merely an assertion.

Thus the Inquiry proceeded without public funding for most participants, with money raised through appeals, art auctions and jumble sales. About £750,000 was spent by objectors but this is not a true reflection of actual costs since almost all those involved reduced or waived their fees. Two of the most well-known parties - FOE and the Council for the Protection of Rural England (CPRE) - raised enough to be legally represented. However, the heavy costs involved prevented them from returning to the Inquiry to follow up new and amended evidence, of which there were enormous amounts. Neither were they able to properly assess the work commissioned by the Inspector from the Cambridge Energy Research Group, following the presentation of the CEGB's economic evidence. In responding to this research as best it could, CPRE reminded the Inquiry that its team of witnesses and advisors was 'dispersed and fully committed elsewhere' and that the Council had 'neither the expertise available nor other resources with which to make a reality of such opportunities for participation' (letter to Inquiry Secretariat, 26/9/84).

Ironically, bodies without legal representation, such as the Town and Country Planning Association (TCPA) were sometimes more able to make return visits, although at considerable personal cost to lay advocates in terms of lost working time, effort and stress. Unable to afford administrative support, lay advocates were also often burdened with such tasks as ensuring that proofs of evidence and reference documents were submitted at the correct time, giving notice of topics for cross-examination, and arranging travel and accommodation. Most were also holding down full-time jobs. Inevitably, these circumstances resulted on occasions in an apparent inefficiency on the part of objectors. Although the Inspector was not unsympathetic to such problems, few adjustments could be made to remedy the gross imbalance of resources. Throughout the Inquiry, the CEGB was served by four barristers, a team of solicitors, several legal assistants and a large administrative unit.

The semi-judicial format

Any discussion of the financial plight of objecting parties at major planning inquiries inevitably turns to the semi-judicial character of such events. Central to this is the barrister who presents the witnesses, 'protects' them throughout cross-examination and re-examines them afterwards to regain lost ground. Within the adversarial system the barrister carries out a selective form of cross-examination - an essentially point-scoring technique designed to undermine the opposition case by demolishing its weaker parts. The training and experience of lawyers prepares them to practice this art in a highly specialised way. This, along with their mannered behaviour and language, sets them apart from the general public at any public inquiry.

At the Sizewell Inquiry the use of this court-like procedure placed objectors in an invidious position. On the one hand, the hiring of counsel and all that it entails is an extremely expensive business, and for almost all of them it was out of the question. On the other hand, a decision to take part in so large and complex an inquiry without legal representation put those participants at a very considerable disadvantage. Although several lay advocates performed well, none were able to match the professional lawyer.

This chapter does not allow for a full assessment of the arguments for and against the use of the semi-judicial format at public inquiries. It is, however, relevant to note that many objectors at the Sizewell Inquiry complained not only about its alien and unwelcoming character but about the inappropriateness of the adversarial system for considering certain areas of evidence. An important feature of the nuclear power debate is its diversity: areas of common concern range from precise technical queries of component design to the enormous moral questions of nuclear waste disposal and plutonium usage. Yet issues like nuclear waste are not well suited to judicial cross-examination: the subjective, philosophical arguments inherent in them call for a more open-ended forum for debate.

An interesting example of this problem was found in the evidence of objectors who came to the

Sizewell Inquiry from Australia and Canada to present evidence on uranium mining. Unlike the CEGB's evidence on the subject, which dealt with future costs and security of supply, these witnesses came to explain the disruption caused by mining works to the local, indigenous ways of life. They spoke of new social problems such as alcoholism, the destruction of ancient sacred burial grounds and other threats to their health and culture. Such 'evidence' was clearly not open to cross-examination in the traditional way and, in these circumstances, the CEGB's decision virtually to ignore it on the inquiry floor was possibly the only reasonable course of action. Inevitably, however, it created the feeling in the objectors' minds that their concerns were deemed to be of no importance, and witnesses who had travelled half way round the world to present their case were understandably bewildered about the nature and function of the inquiry.

This is not to say that the Inspector will not give due weight to the more qualitative issues which - if opinion polls are correct - figure highly in the public's opposition to nuclear power. The point to be noted is that because the material appeared to have a diminished status on the inquiry floor parties felt that they were not getting a 'fair hearing'. The judicial system encourages a situation whereby the value of evidence can be partly measured by the length of cross-examination: the longer a witness remains in the witness box, the more challenging the evidence is thought to be.

Further criticisms were made concerning the inquiry's adversarial format. Several expert witnesses complained that the practice of selective cross-examination was not necessarily effective in identifying the genuine areas of conflict between parties. Scientists in particular also pointed out that the failure of barristers (on both sides) to understand fully the issues at stake was detrimental to a proper exposition of the evidence. Of those who attended the Inquiry as observers, many commented on its excessive formality and alien language. Since it is held as a 'public' inquiry, these seem to be particularly worrying features.

A single inspector

It would be wrong to conclude that no steps were taken to improve the more obvious inadequacies and inequalities of the inquiry process. the Inspector, Sir Frank Layfield, had in fact represented the TCPA at the Windscale Inquiry and was well aware of the problems facing objecting parties. Using the remarkable degree of freedom which an inquiry inspector enjoys, he adapted the system in a number of ways. Taking up a suggestion made by the TCPA when it became known that the Government was unwilling to provide funding, he acquired his own counsel so that some matters which objectors were unable to investigate could nevertheless be pursued. He commissioned his own research on certain economic aspects of the debate. He also invited certain individuals to give evidence. One of these was the South of Scotland Electricity Board (SSEB) chairman who was opposed to the introduction of the PWR. He was willing to listen to evidence on matters which were clearly not regarded by the CEGB as being central to their case: on future waste disposal sites and the hazards of uranium mining. Such actions by the Inspector were welcomed by most objecting parties. They were, however, the actions of an individual inspector: a different inspector could have conducted a much narrower debate, and the breadth of the discussion at Sizewell does not automatically set a precedent for future major inquiries.

The scale and complexity of the Inquiry resulted in widespread agreement that the use of a single inspector was unwise: whatever his or her abilities and experience, no one person should be expected to bear the burden of over two years' evidence and then, alone, to formulate a recommendation on it. Apart from the sheer strain of the exercise, many pointed to the dangers of personal prejudice and the fact that a single person is naturally more open to outside influence in a particular direction than a panel of people, working together.

Conclusions

Even this brief summary of selected procedural aspects of the Sizewell Inquiry (for a fuller account, see Armstrong 1985) indicates that the way in which major and controversial planning proposals are subject to public debate is in need of urgent review. The Sizewell Inquiry fell short of public expectation, and indeed of Government promises, on two major counts: it did not provide a forum for the debate of certain questions which are clearly of utmost importance and concern to the nuclear issue, while its semi-judicial format and unbalanced funding effectively discouraged and often prevented the level of participation which many objecting parties sought. Moreover, there is evidence that the adversarial character of the hearing was inappropriate for much of the qualitative material. The Government's assurance of a 'full and fair' debate proved to be hollow and in some quarters this has served to increase scepticism of the stated desire for public involvement in nuclear decision-making.

Of the major controversies which have been the subject of public inquiries in recent years, nuclear power is clearly one of the most challenging. It brings together an extremely varied band of objectors which, in the light of recent radioactive leakages from Windscale, future plans for waste storage facilities and the proposed plutonium reprocessing plant at Dounreay, promises only to grow. If indeed public acceptability in such matters is as important to the Government as their statements imply, there would seem to be good reasons for admitting that the system, as it now stands, is far from satisfactory.

REFERENCES

ARMSTRONG, J. (1983) The Sizewell Inquiry: a new approach for major public inquiries, London: Town and Country Planning Association, p.149 plus appendices.

Chapter Five

A COMPARATIVE ANALYSIS OF NUCLEAR PLANT REGULATION IN THE US AND UK

John Fernie and Stan Openshaw

This chapter outlines the history of nuclear regulation in the US and UK in order to highlight the differences in approach between these countries in the licensing and planning of nuclear power plants. These topics are of interest because they form an important part, of what some view as a worldwide conspiracy (Falk, 1982) in which successive states (both democratic and otherwise) seek to use nuclear power as a means of centralising and retaining political power. Certainly, it is strange that in Britain the Central Electricity Generating Board (CEGB) openly talks about the need to build US-designed pressurised water reactors (PWRs) on a fairly large scale in the next 20 years: in the US, no order for a PWR has been forthcoming since 1978, with little prospect of any orders in the future. The CEGB claims that it offers superior economics, in the US the PWR has been a massive economic liability to many utilities.

Nuclear plant regulation is also of topical importance because after 40 years of nuclear energy there are growing doubts about the adequacy of the existing regulatory practices in both the US and UK. It seems that current practices are too restrictive and costly for the nuclear industry to tolerate much longer. The industry is anxious for change before the next major nuclear power plant building programmes which are expected in the UK in the late 1990s and in the US in the 21st century. Anti-nuclear groups also view the present as a good opportunity for making the decision-making process more democratic, more responsive to local views, and inclusive of far more stringent environmental constraints.

Certainly, in the US regulatory reform has been on the political agenda since before the Nixon

Administration. A plethora of environmental legislation enacted in the 1960s and 1970s added to the mixture of regulations. This, with various case-specific rulings promulgated by the US Atomic Energy Commission in a context of a federal, multi-agency, administrative system, has created a morass of regulations for the nuclear industry. In addition, the quasi-legalistic nature of the regulatory process has allowed opposition groups to seek redress through the courts and thus delay, almost indefinitely, nuclear plant licensing applications. In the UK the situation is very different but there are also various problems. Until the mid 1970s, two decades after the beginning of civilian nuclear power generation, the whole nuclear decision-making process was conducted in an atmosphere of secrecy, protected from public view by national security considerations. Growing public concern about the scale of the proposed nuclear developments has been translated into various pressure groups, who try to 'open-up' the system of review. This has been achieved mainly through the existing medium of the public inquiry part of the planning process. The 1977 Windscale Inquiry and, especially, the 1983-5 Sizewell 'B' Inquiry provided unprecedented detail of the nuclear regulatory system. However, such inquiries amount to no more than 'talking shops' with no decision-making powers. The Government can simply ignore their recommendations and they tend to be biased towards the nuclear case; for example, the £20-25 million of public money spent by the CEGB in preparing its evidence while no equivalent public funds were available for the anti-nuclear opposition (see Chapter 4 of this volume). The nuclear industry would now like to think that the debate is over and that there is no need for any repeat performances in the future. Nevertheless, the CEGB fears that each proposed new plant may have to face a new public inquiry. Also it is worth noting that in Britain it is the <u>site</u> and not the nuclear reactor that is licensed; hence, the CEGB/SSEB (South of Scotland Electricity Board) want changes to be made so that, as in the US, the generic issues of nuclear safety can be separated from the site specific aspects of power plant development.

In some ways it is ironic that the nuclear agencies in the UK are keen to adopt a US style regulatory approach, whilst the US agencies look towards the British system as having important advantages. It seems that in the absence of outside intervention, a nuclear industry optimum regulatory system would evolve by default. It would be based on the 'best' of both countries, planning systems with the aim of maximising the convenience to the utilities by minimising all possible areas of delay - whether due to public participation in decision-making or to 'misguided' environmental legislation. It is also unlikely that the political context of heightened public awareness will abate, and this factor will act to counter attempts to streamline decision-making. In the UK this could well result in a different form of inquiry system that in practice amounts to a 'steamrollering' exercise. If this situation is to be avoided then it is important that the experiences of the last few decades are not neglected. It is useful, therefore, to examine the history of nuclear regulation in the US and the UK in order to highlight the differences in approach between the two countries to licensing and planning nuclear power plants. Differences in the role afforded to public participation are also important, as are the varying attitudes of the safety authorities towards minimising the residual risks of nuclear power generation. Finally, it will be suggested that one solution to some of the problems might be to adopt the revised licensing system proposed by the Nuclear Regulatory Commission (NRC) in the US as a basis for regulatory reform in the UK.

A REVIEW OF REGULATORY HISTORY

The early years

The development of the civilian nuclear power programmes in the US and UK evolved from the Manhattan Project during World War II. Both countries continued with their nuclear weapons programmes, which resulted in the Atomic Energy Acts of 1946, that established an administrative framework: note that the term 'energy' implies a peaceful intent but in practice it was purely a weapons objective. In the US the legislation gave the newly created Atomic Energy Commission (AEC) unprecedented powers for a government agency, in that it had a monopoly over the use and regulation of all fissionable materials. Although a Congressional committee (the Joint Committee on Atomic Energy, JCAC) was established to oversee the activities of the AEC, in practice its main role was to minimise House and

Senate conflicts during the passage of nuclear related legislation in Congress. These institutional procedures were novel in that they overturned the usual policy making process. Instead of the Executive initiating policy with the consent of Congress, the reverse of this applied to atomic energy.

In the UK also, it seems that atomic energy was given top priority by a nuclear elite that masterminded the weapons programme. Unlike the US, policy making was not vested in the legislature but in a few key figures in the executive; principally Whitehall mandarins, military chiefs of staff, and scientist-engineers involved in wartime R and D projects. The secrecy of the programme was such that apparently many Cabinet members did not know of its extent; it was never debated in Parliament, and the Official Secrets Act and D-Notice system ensured that there was no public or press debate. By the time Churchill returned to power in 1951 he was astounded by the size of the atomic weapons programme (Pringle and Spigelman, 1983).

In the UK public safety was taken into account mainly by the remote siting of facilities and by establishing who was responsible should an accident occur. The administrative organisations in both countries were revised in 1954 with two more Atomic Energy Acts that also started the process of peaceful use of nuclear energy. In the US, the definitive code of federal regulations concerned with nuclear power was outlined, and, with some later amendments, this code has remained the basis of nuclear regulation. The key objective, as outlined to Congress, was to provide 'adequate protection to the health and safety of the public' (Public Law 83-703, 1954; Chap. 14 Sec. 185). The AEC, as an agent of government, was left to determine how best to achieve this objective. It has two modes of reaching decisions; adjudication and rule making. The former is dispute resolution and the latter is a form of policy making. The AEC developed its regulatory rules from case law and adjudications at public hearings. In both instances the decisions made by the AEC can be reviewed by judicial means. In the UK, the 1954 re-organisation transferred responsibility from the Ministry of Supply to a new non-departmental organisation, the Atomic Energy Authority (AEA). In these formative years, the AEA played a dominant role in influencing policy, such as the demographic siting criteria used to site nearly all of Britain's nuclear power stations, while it also developed its own 'in-house' rules and regulations (Marley and Fry, 1955; Openshaw, 1982 a and b). When the first civilian nuclear power programme was announced in 1955, the utilities which would eventually have to use and operate the plants were greatly surprised, since the White Paper (Cmnd 9389) was the first that their engineers knew about the existence of nuclear energy for power generation due to the veil of security that surrounded all nuclear developments (Brown, 1970).

A positive step towards a degree of independence in nuclear regulation occurred after the Windscale reactor accident of 1957. The Nuclear Installations (Licensing and Insurance) Act of 1960 established a system of licensing and inspection to be applied to civilian sites; the AEA was and still is responsible for its own safety. Nevertheless, the Nuclear Installations Inspectorate (NII) became responsible for site approval, the assessment of design safety, and the inspection of plant during construction and operation. Although independent of the AEA and the nuclear industry, nearly all its staff came from the AEA and the nuclear industry; there was no other source of expertise. The 1960 Act also set a limited third party liability of £5 million in the event of an accident. Similar legislation has been enacted in the US in the 1957 Price-Anderson Act which was designed to encourage private utilities to invest in nuclear power and to co-operate with the AEC in its power reactor demonstration programme. The Government indemnity action of the US Act was phased out in 1982 although the UK limited liability still continues today.

From euphoria to uncertainty in the sixties and early seventies

The early 1960s was a period of nuclear optimism. In the US, General Electric and Westinghouse offered fixed price turnkey contracts (they are a complete deal that includes plant, training of staff, and all necessary equipment being supplied by Westinghouse) between 1963 and 1967, proclaiming that nuclear power was competitive with other forms of electricity generation. Utilities, persuaded by the marketing skills of the vendors and having to plan for large increases in demand, placed a flood of orders in the late 1960s and early seventies; see Figure 5.1. In the UK nuclear power was also regarded as competitive with coal fired plant;

Figure 5.1 THE UNITED STATES REACTOR MARKET 1953-1983

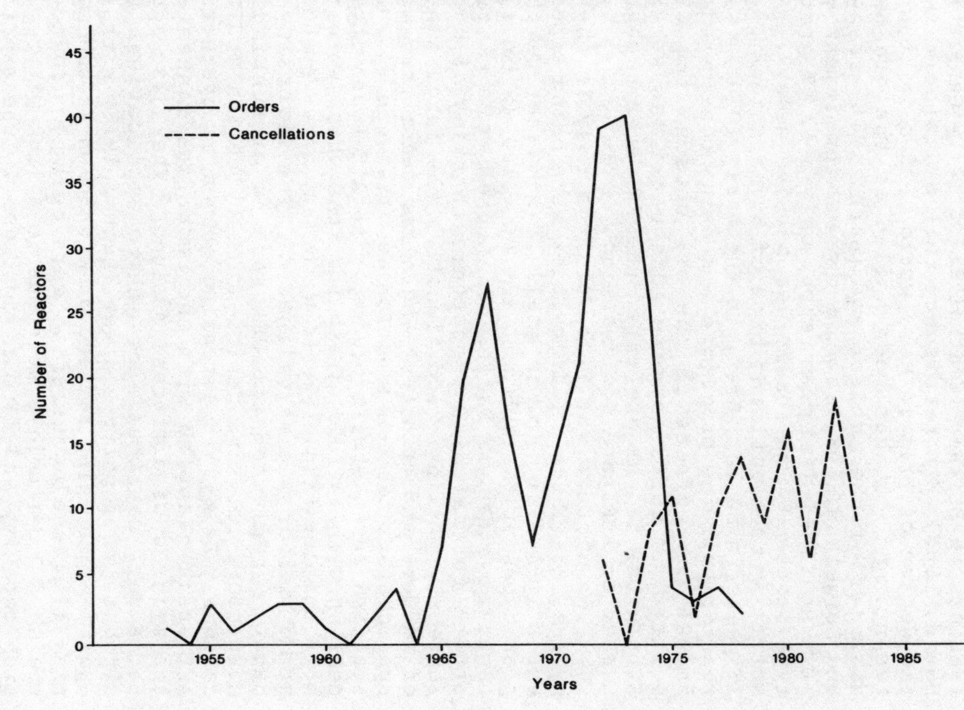

initially because of the value of plutonium produced as a by-product, and subsequently in its own right. The claims being made for the US-designed light water reactors impressed the CEGB, but the continuing influence of the AEA resulted in the selection of the advanced gas-cooled reactor (AGR) as the basis for the second generation of power reactors in Britain (Central Electricity Generating Board, 1965). In retrospect this decision marked the demise of Britain as a world leader in reactor technology. It has been hailed as the greatest disaster of the nuclear age (Burn, 1978; Pringle and Spigelman, 1983). When the AGR commitment was made no final design existed, only a 32 MW prototype reactor. The result was massive delays. Reactors with a building time of 5 years took 10 to 15 years to build. When, in the early 1970s, a new reactor building programme was being considered to offset the effects of the oil crisis, the CEGB refused to build anymore AGRs until one worked. Since then it has preferred the PWR even though a number of AGRs are now 'working'.

The picture was also far from rosy in the US by the late 1960s. Difficulties were being experienced in scaling up the small light water reactor prototypes of the 1950s to the 900-1100 MW (e) size required a decade later. Utilities were placing orders for far bigger reactors than any of those currently working. Additionally, the large number of vendors and architect-engineers meant that most applications for construction permits were for non-standard custom-built plants. The AEC found itself dealing with a flood of applications which were far more complex than hitherto with many site-dependent modifications to be validated. The diversity of designs reflected the US practice of seeking to compensate for site deficiencies by design modifications. Delays in licensing ensued as overworked AEC staff grappled with the technical issues and the myriads of different designs. They also adopted a very cautious approach to applications for sites in metropolitan areas (Bunch, 1978) with all applications being refused or withdrawn at the same time that analogous sites were being approved in the UK. Other delays were due to a lack of production capacity and labour shortages in the specialised construction trades. Nonetheless, between 1967 and 1973, the average delay per plant was only 5 months (Quirk and Terasawa, 1981).

The 1960s also witnessed the rise of populist environmentalism in the US. This commanded political support and ultimately spawned a plethora of legislation which was to influence utility decision-making. Interveners in the licensing process, however, only played a marginal role in delaying projects. For example, from 1962 to 1971 only 27 out of 100 construction permits were contested in public hearings and only 15 cases went to judicial review (Rolph, 1979). It is interesting that the equivalent nuclear 'honeymoon' lasted until the late 1970s in the UK. Nevertheless, the seeds of doubt about the safety of nuclear technology had been cast and far more coherent opposition to nuclear power started to plague the power utilities in the 1970s and 1980s. Initially, most of these challenges were environmental in nature. Thermal pollution of rivers was a key issue and the state governments and local pressure groups were the main interveners. However, after the National Environmental Policy Act (NEPA) of 1969 and the Calvert Cliffs judicial decision of 1971, which forced the AEC to include an environmental impact statement in the licensing procedure, this institutionalisation of environmental protection led to a shift in public attention from environmental towards safety issues in the 1970s (Kasperson et al., 1980).

The question of reactor safety was being increasingly raised by interveners in response to the discovery that the AEC staff themselves had doubts about the various untried emergency core cooling systems (ECCS) which were being used by reactor manufacturers to allow their designs to meet existing safety standards. For example, a construction permit was given for Indian Point 2 in 1966 with serious reservations being expressed about the proposed ECCS (Gillette, 1972). The AEC convened a rule-making hearing in January 1972 in order to try and defuse fears on the ECCS issue and speed up the licensing process which had been slowed by interveners objecting on similar grounds at different plants. The hearing was intended to last six months but it continued for 18 months. It soon became clear that many AEC employees were also questioning the reliability of the ECCS. Interveners, armed with 'leaked' material and other information procured through the Freedom of Information Act, were able to show that the AEC had overcommitted resources to advanced reactor systems at the expense of improving light water reactor safety. Nuclear critics attacked the AEC for stressing promotion rather than regulation. The Legislature was quick to respond and the Energy

Organisation Act (1974) disbanded the AEC and created two separate agencies; the Energy Research and Development Administration and the Nuclear Regulatory Commission (NRC).

The post-oil crisis era

The latest period of nuclear history has been a traumatic one for the electric utilities in the US. Market conditions became increasingly unfavourable with a stabilisation of demand, an excess of capacity, and large financial commitments to capital intensive projects on order or under construction that generated no income. Electricity consumption fell to zero growth and then declined in the early 1980s, when utilities were planning 10 years ahead on the assumption that demand would grow at between 5 and 10 per cent compound per annum. With falling profits, stock values plummeted. Utilities began to cut their losses and started cancelling nuclear projects. Between 1972 and 1984, 116 reactors were cancelled (UCS, 1984; USDOE, 1984); see Figure 5.1. Even nearly-complete plants were scrapped because of the cost of completion and of back-fitting to meet recently changed safety standards.

The plight of the US nuclear industry has often been attributed to the regulatory environment (Gordon, 1984; Starr and Braun, 1984) but regulators (Ahearne, 1984; Palladino, 1981 and 1984) and others (Quirk and Terasawa, 1981; US Office of Technology Assessment, 1984) have pointed out that delays in commissioning and the associated financial difficulties can be attributed to weaknesses within the industry itself rather than to regulatory constraints. It is true that utilities have had to comply with a growing list of regulations but the objectives of new legislation were to make plants safer, to minimise their environmental impact and to keep electricity charges to consumers at acceptable levels. Not only did the utilities have to adjust to the environmental legislation but also to increased State involvement in nuclear policy making. State Energy Departments were created and, in liason with Public Utility Commissions (PUCs), began to subject utility investment plans to greater scrutiny. Certain States, notably California and New York, began to place more emphasis on 'soft' energy paths. Utilities building nuclear plants began to request electricity price increases for work in progress but the PUCs either were reluctant to allow this or else they reduced the rate of increase to exclude costs that could be attributed to management inefficiency. Jaworksi and Ward (1984) give a detailed account of the response by Michigan Public Service Commission to a 968 million dollar rate request from Detroit Edison.

In addition to growing financial problems, the power utilities faced a far more co-ordinated opposition to their plans after 1973. Whereas interveners had previously been a fragmented assemblage of local environmental groups objecting to specific proposed plants in their areas, the 1970s saw the rise of national coalitions of environmental groups with an anti-nuclear strategy. Nader's Critical Mass, the Union of Concerned Scientists, and the Sierra Club were at the forefront of this crusade. They politicised the nuclear issue, mobilised public opinion and secured a degree of support from political figures at State and Federal level. They also used the medium of 'legal advocacy' (Cook, 1980) to draw attention to their case. This has been facilitated by the liberalisation of the 'doctrine of standing' which defines those who can sue or act as interveners in public hearings. Initially, only locally-affected persons or agencies could do this, but by the 1960s environment groups with local representation but national resources and organisational structures were allowed to intervene. In general it seems that the judiciary dismissed the vast majority of cases brought against the nuclear industry; the principal exceptions being the Calvert Cliffs (1971) decision and the Supreme Court's (1983) decision to uphold the State of California's moratorium on nuclear plant. Nevertheless, by resorting to the courts, nuclear opponents can effectively delay commissioning by a long drawn out appeals process.

In some ways the environmentalists were vindicated in 1979 when the Three Mile Island accident showed that accidents could occur beyond the design basis. Criticisms of both the industry and the NRC had severe ramifications for the future of nuclear power in the US. The Kemeny Report (Kemeny, 1980) made a number of recommendations which led to urgent action by the NRC (NRC, 1980). Utilities had now to consider accidents beyond the design basis (that is accidents with which the design could not cope), safety reviews and a major research programme started to consider how best the residual risk of nuclear power could be reduced by siting and more thorough safety assessments (Larkins and Cunningham, 1983; Fernie and Openshaw, 1985).

The nuclear industry is undergoing a crisis of confidence. Although the Reagan Administration strongly advocates nuclear expansion (Katz, 1984), utilities have not shared this commitment. Starr and Braun (1984) feel that many of the problems relate to the fragmentary nature of the industry, with hundreds of utilities, four reactor vendors, and a large number of competing architect-engineers. This fragmentation has reduced the opportunities to learn by experience, while differential standards and poor performance have attracted bad publicity for the whole industry. To resolve some of these problems, the Atomic Industrial Forum recommends that consortia of manufacturers and architect-engineers should be formed, based around the most experienced companies, to improve upon existing ad hoc organisational arrangements whereby utilities commission custom-built plant from a shopping list of companies (Walske, 1984).

In the UK similar organisational difficulties had been evident in the 1960s and had led to the emergence of a single nuclear power construction company. The construction difficulties with the AGR resulted in the National Nuclear Corporation (NNC) being created in 1973 as the sole architect-engineer-construction company to build plants for the two State utilities (i.e. CEGB and SSEB). Since then the principal obstacle to a long term nuclear strategy for the nationalised State power utilities has been the thorny issue of reactor choice. Although the CEGB declared a preference for the PWR in 1973, various factions within the nuclear industry, principally the AEA, wished to remain loyal to British technology; first with the abortive steam generating heavy water reactor (SGHWR) and then latterly the AGR again. The result of the Sizewell public inquiry will probably finally resolve this issue for the CEGB, even if the SSEB still strongly favours the AGR, and the AEA is starting to put up a strong case for a commercial fast breeder reactor (FBR). The 1970s was a period of uncertainty about reactor choice at a time when, however, the urgency for a major reactor construction programme lessened. The 1980s will probably see the former problem resolved and the beginnings of a large reactor building period as the utilities seek to reduce their dependence on coal and move towards a total dependency on the nuclear option before opposition becomes too great to handle.

One of the main differences between the US and UK has been the lack of success of environmental groups in influencing UK policy. It is true that a number of pressure groups have grown in importance over the last decade but, unlike their US counterparts, they are restricted to rigid political processes through which to channel their views; for example, public inquiries and evidence to Parliamentary committees, neither of which are particularly effective. Informal channels of consultation do exist but these are largely dominated by a powerful pro-nuclear lobby (Lowe and Goyder, 1983). Furthermore, UK anti-nuclear groups have no equivalents to the US Freedom of Information Act to which they can turn, and no rights of legal redress. They have to rely on 'leaked' documents from Government Departments to gain an insight into the policy making process, or details of reactor safety reports; and even then nearly all the useful information is protected by the Official Secrets Act. In the US, Government agencies can be taken to court for breaking their statutory brief; a tactic used by environmental pressure groups (Cook, 1980). Similar statutory obligations exist in the UK but they are considered non-enforceable by law (Fernie and Openshaw, 1984; Macrory, 1982). There is simply no right of appeal against many nuclear decisions (Purdue, Kemp, and O'Riordan, 1984).

US - UK historical comparisons

It is clear from this discussion that nuclear power has failed to live up to the high expectations of the early post-war period in both countries. The course of a new technology which was granted elevated status in political decision-making in the 1940s, 1950s and 1960s began to wane in the ensuing decades with changes in the political and economic environment. The mid-1960s was the watershed period in the history of nuclear power. In the US, turnkey contracts led utilities to order plants on a massive scale, stretching the licensing system and revealing technological problems in scaling up prototypes to commercial size reactors. This coincided with the rise of environmentalism and the use of the legal system by opponents of nuclear power to cause delays and abandonment of new projects. Even before Three Mile Island, utilities were cancelling plants and new orders were not forthcoming as costs soared, and revenues remained depressed in the wake of the energy price shocks of the 1970s. In the UK, problems began in 1964 with the decision to order AGRs for the next stage of the nuclear power programme. Once again, expectations were not fulfilled and reactor choice became a contentious political football throughout the 1970s and is still not resolved. However, in the UK the State power utilities are not as sensitive to market forces, and they have been able to hide effectively the real costs of the AGR delays. They have a monopoly of the market, can set their own prices, are able to make low interest loans, and can even claim that an AGR which is over 12 years late is still economic. In the US such a reactor would have been abandoned. In the UK, the costs of alternative fossil fuel replacement power are ignored - unless, of course, they are used to show the adverse effects of any further delay to the Sizewell 'B' station. The exclusiveness of UK nuclear decisions has meant that there have been virtually no public reviews of policy. There have been no real debates in Parliament and information has been refused to select committees charged with investigating certain aspects of nuclear power. Nuclear opponents have formed into various public pressure groups but they have virtually no power. They are probably infiltrated by Government security agencies, they are certainly fed with a considerable amount of misinformation, they are under continuous surveillance as potential subversive organisations, and they have no legal powers to interfere.

The only area of success has been in obtaining a degree of 'opening up' of the debate. The Flowers Report (Flowers, 1976) had reservations about a wholehearted commitment to nuclear power until certain problems had been resolved. It acted as a catalyst to greater public debate at a time when the Energy Secretary was advocating a greater openness on nuclear issues and, in an unprecedented move, asked the AEA for answers to a whole series of questions put to him by anti-nuclear groups. These trends have continued with the publication of a Monopolies and Mergers Commission (1981) report and various parliamentary committee reports; for example, the House of Commons Select Committee on Energy (HCSCE, 1981) and House of Lords Select Committee on the European Communities (HLSCEC, 1984), both of which were openly critical of the CEGB's nuclear programme. The Sizewell 'B' public inquiry is another landmark in UK regulatory history. It is the first time that both the licensing and the planning consent process have been simultaneously publicly examined, although licensing aspects were not properly covered and the pro-nuclear lobby considered the result to be a foregone conclusion.

THE LICENSING PROCESS

The US approach

In order to achieve a better understanding of the review process in the US and UK it is relevant to discuss the procedures involved in securing a licence or consent to operate a nuclear plant in both countries. In both cases the overall requirements are similar; utilities are obliged to apply for permits to comply with a large number of regulations on matters which range from the abstraction of cooling water to pollution control measures to the storage and transport of radioactive materials to and from the site. The first obstacle is to obtain a licence for site and design approval.

In the US, the licensing process has evolved from a narrow dialogue between the AEC and the applicants in the early 1960s, to an extensive review of anti-trust and safety and environmental issues in the confrontational atmosphere of the 1970s and 1980s. The basic sequence of stages is shown in Table 5.1.

Table 5.1 CURRENT AND PROPOSED LICENSING PROCESS FOR US NUCLEAR POWER PLANTS

Stage	Activities
Current process	
1a	application for a construction permit covering design and safety features and an environmental review of the proposed development.
1b	copies of the application are available for public inspection.

Table 5.1 (continued)

Stage	Activities
Current process	
2a	Anti-trust review of licence application by the NRC and Attorney General.
2b	NRC reviews the application and submits a draft environmental statement and a safety evaluation report (SER) for agency, interest groups and public comment.
3	Final Environmental Impact Statement is prepared incorporating these comments.
4	The Advisory Committee on Reactor Safeguards (ACRS) reviews the application and the NRC issues a supplement to the SER.
5	The Atomic Safety and Licensing Board (ASLB) holds a mandatory public hearing to decide on the application.
6	The ASLB grants or denies a permit; the decision is subject to appeal.
7	Application for an operating licence is made a few years before completion and the applicant submits further, updated, evidence in support of the application.
8	NRC and ACRS review the application; a final safety evaluation report and an updated environmental statement are issued.
9	A public hearing (not mandatory) may be held.
10	NRC maintains surveillance of the plant throughout its operating life.
NRC proposals for a new streamlined process	
1	Early Site Review Procedure: site use permit application reviewed by NRC and other agency and interest groups. Approval for a 10 year period. Possible public hearing.

Table 5.1 (continued)

Stage	Activities
2	Design Certification Procedure: in liaison with 1, standard design of plant will be certified for 10 years. Possible public hearing.
3	Combined Construction and Operating Licence Procedure: as a result of 1 and 2 considerable streamlining of existing process takes place. Public hearing.
4	Limited public hearing before start up.
5	NRC maintains surveillance of the plant throughout its operating life.

Public participation is a feature of the nuclear licensing process that has been instrumental in influencing Government policy and regulatory rule making. The barriers of secrecy began to break down in 1957 when a construction permit for a fast breeder reactor (Fermi 1) near Detroit was the subject of dispute between the Advisory Committee on Reactor Safeguards (ACRS) and the AEC. The ACRS recommended that the permit should be denied; the AEC disagreed and granted the permit. One of the commissioners who agreed with the ACRS released the ACRS report, which at the time was considered to be confidential information. The furore which followed meant that Congress amended the 1954 Act to provide public access to ACRS reports and mandatory public hearings before a construction permit could be issued. Since then, legislation on anti-trust matters (1970), on NEPA (1969), and on Freedom of Information (1966, 1974, 1976) have facilitated public input into the licensing process.

This process, with some of the regulations, has been criticised for being complex, and for duplicating the functions of different agencies. Critics have argued a need for regulatory reform (Calzonetti, 1981; Kemeny, 1980; Gordon, 1984). Indeed the licensing process is now under review and a streamlined process (Table 5.1) will probably soon be developed, even if this is largely regarded as an academic exercise because no new applications are now expected in the foreseeable future. The expected changes will involve advance approval of standard designs and sites, and a one-step construction and operating licence procedure. The hope is that the siting and design reviews will allow the public to influence decisions <u>before</u> utilities commit resources for constructing a plant on a particular site. Instead of wide-ranging public hearings, attention will be restricted to operational issues such as emergency planning procedures.

Environmental groups have been hostile towards these proposed changes as they would limit their opportunities to discuss overall nuclear policy and would confine debate primarily to issues of siting and safety. In essence, 'fishing expeditions' by interveners attempting to stall the licensing process will be eradicated. On the other hand it can be argued that there are already sufficient nuclear sites in the US to meet the needs of the nuclear industry in the 21st century and that in the next round of reactor building the principal concern will be the redevelopment and extension of existing sites, rather than the development of large numbers of new greenfield sites. If this happens, then the opportunities of anti-nuclear groups to oppose nuclear developments will anyway be far more restricted than previously was the case.

The UK approach

The recommendations in the US of a one-step licensing process is a move towards the UK system of nuclear power plant approval. The main difference between what is proposed in the US and what exists in the UK is the limited level of public input into the process in the UK. Table 5.2 outlines what is involved at present.

Table 5.2 THE LICENSING AND CONSENTS PROCESS FOR UK NUCLEAR POWER PLANT

Stage	**Activities**
1	Pre-application informal discussions with various Government Departments and local authorities about possible sites. Leads to publication of a few areas of interest where drilling is required to confirm their technical suitability.
2	A site is adopted for a nuclear power station by the CEGB (or SSEB) although no application

Table 5.2 (continued)

Stage	Activities
	has been submitted as yet and no firm commitment to power station development made.
3	An application for planning permission to build a nuclear plant and obtain a nuclear site licence. The major decision has now been made by the utility to develop the site although no public information review or any information has been provided.
4	If the Local Authority is in favour of the development then, provided the Nuclear Installations Inspectorate approves the safety aspect of the design, the site will be developed once the Government gives loan sanction.
5	If the Local Authority opposes the development, then there will be a public inquiry; the terms of reference of which are established by the Secretary of State for the Environment. He may also call a public inquiry as is the tradition for the first application for a reactor not previously built in the UK.
6	If there is a public inquiry, then the Inspector's recommendations are considered by Government who may or may not agree with the suggestions. The final decision, as in 4, will be made by the Cabinet. The public inquiry is an information providing rather than a decision-making body.

The utility (i.e. the CEGB or SSEB) must secure consent from the Department of Energy to comply with section 2 of the Electricity Lighting Act (1909), sections 33 and 34 of the Electricity Act (1957), and secure a licence from the Nuclear Installations Inspectorate (NII) according to the Nuclear Installations Act (1965). By the time a formal application is made for planning permission - and this is the first time that either the CEGB or the SSEB will admit to having a definite intent to develop a nuclear power station at a particular location - the chances of influencing the decision are minimal; see stage 3 in Table 5.2. Possibly by stage 1 and certainly by stage 2, the decision will have been taken; the utility will have a preferred site in mind, it will have probably completed detailed site investigations and even detailed engineering design work. Local opposition will be noted but it is not considered relevant because of the overriding significance of the national interest arguments. In recent years the CEGB has tried to influence public opinion by large scale public relations exercises in areas under investigation but there is no formal public participation component to the regulatory process, and this makes it quite different from most other areas of planning. Indeed it seems that public review of the detailed aspects of nuclear planning is definitely not encouraged for fear of the knowledge of what is involved, especially in relation to nuclear safety, will result in problems for the utilities. As a result the evaluation of alternative sites is usually done in secret, probably very informally, and is complete before the end of stage 1 (Openshaw, 1986). Once a written application is made, it is for the development of a particular site. It is usually concerned - and the utilities would always like it to be - with local site specific aspects (for example, landscaping and car parking arrangements) but excludes virtually all issues related to need, safety, and national nuclear policy matters. The only exception so far has been the Sizewell Inquiry and this is recognised to have been exceptional and unlikely to be repeated.

A public hearing may not be necessary if the local planning authority approves the project and grants planning permission (stage 4), unless the Secretary of State calls for one. In some cases, local authorities will have reservations when initially consulted and the utility will make further attempts to gain a positive response. In the past a kind of horse-trading has been involved, whereby a utility informally agrees to fully develop one site before moving to another in an area which may be considered more sensitive. Increasingly, local authorities are refusing to co-operate, especially if the utility does not consult them fully about their intentions. At one site (Druridge Bay), for example, the CEGB has bought land and virtually completed detailed design work while continuing to maintain that it has no commitment to any development. When negotiations break down there will be a public inquiry which may be restricted to local planning aspects, but where various interest groups, not party to any of the informal negotiations, will have an opportunity to air their feelings. Public inquiries are not, however, decision making bodies: only a means of ensuring that the Minister and Cabinet are better informed before projects are approved. The utility would normally expect that it would be allowed to proceed; indeed whilst the Sizewell 'B' Inquiry was in progress the CEGB actually ordered £100 million of PWR components, despite having no consents either to build or operate a PWR in Britain.

Of the twenty-two planning applications made for nuclear power stations, only eleven have been subject to a public hearing (Owens, Hills, and Cope, 1983). The Torness nuclear power station inquiry lasted less than one day in the early 1970s. A decade later permission to erect a railhalt for waste transport was refused and the public inquiry lasted three weeks. But, once again, the greater public involvement in nuclear matters operates only as a delay mechanism. In Britain the effects are minimal because the additional delay costs are readily absorbed. Furthermore, the history of public inquiries into nuclear power applications shows that the inspectors have invariably excluded issues not closely related to the applications - for example, safety or need for security - because they are outside the terms of reference of the planning inquiry system (Wynne, 1982). It seems that virtually all avenues of possible objection on safety and environmental grounds are protected from public scrutiny. Whereas US utilities have to submit volumes of safety and environmental data from NRC and public review, the UK utilities are not even required to submit an environmental impact assessment and they have no need to publicly justify the sites they select. Additionally, it is now known that in Britain the impact of power station developments on the environment has not been considered to be a relevant factor in the choice of existing sites.

Even if the UK adopts the EEC-style environment impact assessment, it is not thought that this will bring about any changes. The CEGB would claim that its current investigations are thorough enough for this additional statutory requirement to present no problem. Similarly, the question of safety has not been investigated by bodies outside government research agencies. Most safety documents are classified; indeed during the Sizewell Inquiry some documents had to be acquired through the Freedom of Information Act in the US because they are not available in the UK. The NII has not published any detailed safety assessment of either reactor or site during the inquiry. It is now accepted that its final report will not be available until after the decision about the planning application will have been made. Compared with the openness of the work of the US NRC for public consultation, the UK NII is extremely secretive. Its first annual report was published only in 1978, 18 years after its creation. Moreover, its problems are compounded by a shortage of staff and the need to augment its workforce with personnel seconded from industry, including the CEGB! This element of agency capture has meant that the safety problems have been played down and details are not revealed to the public. It must be doubtful whether NII has performed comprehensive in-depth safety studies of all the nuclear plants currently operating in the UK, and this may partly explain its reluctance to publish details.

The Sizewell 'B' Inquiry was somewhat exceptional in that the terms of reference were deliberately broadened to allow a full airing of all issues of public interest. This is an improvement, albeit a minor one, because it is a traditional practice in the UK to hold a full inquiry for the first reactor of a new type. The utilities expect that this one-off inquiry will settle once and for all the generic issues concerning need, cost, safety, and design aspects so that subsequent PWR applications can be restricted to local planning matters.

There is now considerable debate about the use of the public inquiry as a tool in the large national interest projects and much more discussion on how it can be improved (Outer Circle Policy Unit, 1979; Rowan-Robinson and Edwards, 1981; Pearce _et al_., 1979; Macrory, 1982; Armstrong, 1985). Table 5.2 shows that the public inquiry comes at the end of the licensing process. Effective participation is limited because the decision about the site and the need for the plant has been already taken before the full consultation in public. Many of the critics of the present system advocate a multi-stage inquiry procedure to separate generic from site-specific issues. The Department of Environment can invoke the Planning Inquiry Commission: this has been on the statute books since 1968. However, successive governments have been unwilling to implement its powers, namely: to review proposals, carry out investigatory research, consider alternative sites, and then conduct one or more site-specific local inquiries. Governments are naturally cautious about any procedure which opens up their own policy to greater public scrutiny, so in practice nothing more than tinkering with the existing public inquiry machinery has occurred.

REGULATORY REFORM

Whatever new institutional arrangements may be proposed, British nuclear regulators ought to follow the lead of their counterparts in the US. Nuclear power is regarded as being an essential evil for the future of mankind. Its success is a fundamental prerequisite for high living standards in the latter part of the 21st century, perhaps even before. It is important that this long term commitment to an all-nuclear future is acceptable to the majority of the population, otherwise public safety and vast amounts of public money could be at risk. It is imperative, therefore, that the longevity of nuclear sites is recognised and allowed for when planning the next phase of nuclear reactor building. This requires a far more open debate, greater public participation, and a far more realistic attitude to be adopted by the power utilities. Public acceptability now, and for evermore, is a fundamental necessity and cannot be guaranteed in a democracy by dictatorial methods. In order to restore public confidence in nuclear power utilities and to meet the goal of public acceptability, a modified form of licensing is urgently required.

In the US the NRC intends to streamline and improve safety standards. This was to be achieved by obtaining Congressional acceptance for pre-approved standard designs and sites. Standard designs would enable the NRC to monitor plant performance more effectively than hitherto and a revised siting policy would almost certainly propose remote areas with a well-established, existing, nuclear infrastructure for future development. This strategy would cause 'minimum stress on existing institutional structures' (Burwell et al., 1979; p. 1044) and comply with the NRC policy of seeking to minimise the residual risks of a nuclear accident. Residual risks or the effects of large scale but highly improbable accidents now dominate the nuclear risk assessment process; because the consequences of anything other than a 'big' accident are considered to be catered for by existing designs and are therefore considered to be zero.

Even before Three Mile Island, a siting task-force was established to accomplish the following goals:

1. to strengthen siting as a factor in 'defence-in-depth' (if all else fails remoteness will reduce the consequences a little) by establishing requirements for site approval that are independent of design considerations;

2. in site assessment, to take into consideration the risk associated with accidents that are beyond the design basis (so-called class 9 accidents) through the use of population density and distribution criteria to limit the magnitude of any health consequences;

3. to select sites that minimise the risk from energy generation (NRC, 1979).

It should be stressed that the defence-in-depth strategy has long been traditional in the US. The 1962 reactor siting guidelines (USAEC, 1962) included a minimum separation distance from the reactor to the nearest population centre of 25,000 or more people as a last line of defence from class 9 accidents. However, applicants for sites did not have to consider the possibilty of such an unlikely event actually happening until after Three Mile Island. Currently, the question of precisely what specific population criteria are recommended is in abeyance pending the results of research into new categorisation of radioactive releases from a reactor accident and the implementation of a severe accident policy. Apparently, the amount of radiation released at Three Mile Island was considerably less than forecast due to greater retention by the internal surfaces. This fact has led to a major review of estimates of release by a set of standard accident scenarios used in safety studies.

The CEGB is also keen to obtain pre-approval of sites because this would facilitate long-term planning, regardless of government indecision over the choice of reactor or energy policy. Baker (1982) adds that:

> the suitability of a given site for power station development, including its suitability

> in relation to environmental impact, could be established independently of, and in advance of, discussions on the question of need. In this way, a small 'pool' of sites, cleared as technically and environmentally suitable for development of a range of generating systems could be built up and from these options a selection could subsequently be made as required to meet the needs and policies of the time (p. 16).

He does not point out that the existing nuclear sites constitute an effective pool, in that it is the CEGB's intention both to redevelop them and extend them as necessary. Many of these activities would not require planning permission.

In principle the site 'pool' approach is commendable but the CEGB and the SSEB would have to be far more open about their siting strategies and site evaluation procedures than in the past. Whereas the entire land area of the US has been thoroughly screened for site suitability for nuclear developments using computer automated methods, and suitable data bases have been established, this is not the case in the UK (Fernie and Openshaw, 1985; Openshaw, 1986). The CEGB consistently argues that sites are chosen after a thorough investigation, that existing consultation procedures are appropriate checks on its proposals, and that environmental impact statements are unnecessary (Baker, 1982; Howells and Gammon, 1984). In practice, the Board evaluates only a few sites (two or three) and the entire process is secretive. When its siting strategy was investigated at Sizewell, it became apparent that the decision to choose Sizewell was first made in principle (in 1978) without any comparison of sites, and then justified retrospectively (three years later) when a public inquiry was imminent.

The relative complacency shown by the CEGB over site selection, compared with its US counterparts, can be explained by the different policies pursued by the NII and NRC towards the problem of minimising the residual risks to the public who live in the region of a nuclear power plant. The NRC is strengthening its already existing policy to take into account unlikely accidents, but in the UK siting has always been regarded as of only marginal relevance to the safety debate. The view is expressed that since it is impracticable to formulate a siting policy that would have any great effect on the consequences of a 'big' accident, then the only satisfactory solution is to ensure that such accidents can never occur. This is clearly the view of the nuclear establishment, and its incorporation into radiological protection stands developed by supposedly 'independent' agencies is somewhat worrying. It is leading to a situation where the same US reactor can be built in both the UK and US; yet in the two countries it would be subject to very different siting policies. In the US, with years of experience of PWR design and operation, siting would be used as a design-independent safety measure to minimise the effects of improbable big accidents. In the UK, with no experience of PWR construction or design or operation, such measures are considered inappropriate. It seems that the safety of the British public is to be placed in the hands of not-yet tested, let alone developed, microprocessor-based safety control systems.

Openshaw (1982a, b; 1986) shows that 'better' sites do exist in the UK than many of those currently in use by the CEGB and SSEB. If the object is to minimise the residual risks of nuclear power then it is also possible to adopt US-style siting criteria without precluding the nuclear option from many parts of Britain. Moreover, if the aim was to maximise the likely long-term public acceptability of nuclear power by adopting a 'real' remote siting policy, then this can also be done in the UK. The current policy of using existing and adopted, but not yet developed, sites (and this could include current conventional power station sites) leads to inconsistencies. For example, it is interesting to note that sites currently nominated for PWRs have widely different demographic characteristics. Sizewell and Dungeness are the most 'remote' whilst Druridge Bay (a greenfield location adopted in 1982) has demographic characteristics which would have certainly precluded it from development if it was proposed in the US. Likewise, the Hartlepool and Heysham sites would never have been permitted in the USA at all.

If the regulatory reforms earmarked for the US were modified to British planning practices, the main focus of a local planning inquiry would relate to emergency planning procedures. Once again, however, the ultra-cautious approach of the US safety authorities and their acceptance of the possibility, however remote, of a class 9 accident (an accident beyond that which the design safety measures can handle) has provoked a different response to emergency planning than that adopted in the UK. The CEGB has undertaken only limited studies of the possible consequences of accidents beyond the design basis; even then, it has assumed that the built-in engineered safeguards would either prevent the release of radioactivity or contain it to a narrow geographical area (George and Hilsley, 1982). This optimism is reflected in UK emergency planning procedures which assume that people would need to be evacuated for a distance of only two miles from the reactor in a worst-case accident; indeed this seems more appropriate for a fuel flask or spent fuel pond accident than anything more severe. Similar evacuation measures had been in force in the US until the Three Mile Island accident. Since then, revised emergency planning zones have been designated with a 10 mile evacuation zone and a 50 mile food ingestion control area (NRC, 1981). Even these revised regulations have been criticised on the grounds that the evacuation zone is too narrow in terms of the response by residents to a real, rather than an hypothetical accident (Zeigler and Johnston, 1984 and Chapter 12 of this volume; Cutter, 1984). This was the case at Three Mile Island, the only major nuclear emergency in the West, where evacuation behaviour was unpredicted, with a massive over-response by residents to advice from the emergency agencies. In such a crisis, these agencies were much less coordinated than they had been seven years earlier in tackling Hurricane Agnes; a hazard with a proven track record (Flynn and Chalmers, 1980).

In the UK, however, it is assumed that 'an accident at a nuclear plant could present a different type of hazard but it is highly unlikely to pose a significant threat to public safety or require actions to protect the public which are different from those required for other civil emergencies' (HSE, 1982; p. 1). This is, of course, incorrect. Nevertheless, this attitude prevails; at the Sizewell Inquiry, the emergency services' personnel exuded confidence on their ability to cope in a crisis. It is interesting to note that Cutter (1984) expresses concern about three sites in the US (out of ninety-one with operating licences or construction permits) because it would be impossible to evacuate individuals if a major accident occurred. Her criterion for these three sites was that there was 100,000 people living within ten miles of the reactors. In the UK, four out of the eighteen sites also have this demographic characteristic. Three of these sites are considered to be 'remotely' located with MAGNOX reactors (Bradwell, Berkeley, and Oldbury). One of the semi-urban sites (Hartlepool) falls into this category while the only new site currently being proposed (Druridge Bay) has 97,000 people within 10 miles of it; a much higher figure than most other existing sites.

Conclusions

Regulatory reform should be a priority issue in both the US and UK. A two-tier system of review seems to be sensible. The generic issues of national importance should be the subject of specific national and state policy objectives and cannot really be left to utilities to sort out. In the UK this would imply that central government would take on the responsibility for safety review and designate sites from which the utilities could then choose. The local inquiry would deal with matters of local relevancy including emergency preparedness. We advocate that the UK safety authorities should adhere more closely to the revised licensing system that the NRC has recommended to Congress. This would encourage far greater public participation at a much earlier stage than at present. Indeed if PWR approval is given at Sizewell, then the UK safety authorities should review their siting and safety philosophy, which is now more than 16 years old and more relevant to gas-cooled reactors with their greater levels of intrinsic safety. A more cautious approach, such as that adopted by the NRC, is advisable to minimise the residual risks of nuclear power generation and to ensure maximum acceptability for nuclear power both at present, after Chernobyl, and for the future when, maybe, several serious accidents will have happened worldwide. The longevity and the irreversibility of many of the nuclear decisions makes the switch to a more publicly defensible licensing system absolutely essential, if nuclear power is to have a future in a democracy. The present time offers an excellent opportunity to meet this challenge and develop appropriate regulatory systems for the future.

REFERENCES

AHEARNE, J. F. (1984) 'US nuclear power - has its time passed?', interview in Resources 77, 5-8.

ARMSTRONG, J. (1985) The Sizewell Report: A New Approach for Major Public Inquiries, London: Town and Country Planning Association.

BAKER, J. (1982) 'The legal framework for nuclear development', in R. MACRORY (ed.) Commercial Nuclear Power, London: Imperial College.

BROWN, S. (1970) 'Cockcroft Memorial Lecture: The background to the nuclear development programme', Journal of the British Nuclear Energy Society, 9, 4-10.

BUNCH, D. F. (1978) Metropolitan Siting; A Historical Perspective, NUREG 0478, Washington DC: Nuclear Regulatory Commission.

BURN, D. (1978) Nuclear Power and the Energy Crisis, London: MacMillan.

BURWELL, C. C., OHANIAN, M. J. and WEINBERG, A. M. (1979) 'A siting policy for an acceptable nuclear future', Science 207, 1043-1051.

CALZONETTI, F. J. (1981) Finding a Place for Energy: Siting Coal Conversion Facilities, Resource Papers in Geography, Washington DC: Association of American Geographers.

CENTRAL ELECTRICITY GENERATING BOARD (1965) An Appraisal of the Technical and Economic Aspects of Dungeness B Nuclear Power Station, London: CEGB.

COOK, C. E. (1980) Nuclear Power and Legal Advocacy, Toronto: Lexington.

CUTTER, S. L. (1984) 'Emergency preparedness and planning for nuclear power plant accidents', Applied Geography 4, 235-245.

FALK, J. (1982) Global Fission: The Battle over Nuclear Power, Oxford: Oxford University Press.

FERNIE, J. and OPENSHAW, S. (1984) 'Policymaking and safety issues in the development of nuclear power in the United Kingdom', in M. J. PASQUALETTI and K. D. PIJAWKA, (eds.) Nuclear Power: Assessing and Managing Hazardous Technology, Boulder: Westview.

FERNIE, J. and OPENSHAW, S. (1985) 'Nuclear power in the US and UK: the role of siting in safety philosophy', in F. J. CALZONETTI and B. D. SOLOMON, (eds.) Geographical Dimensions of Energy Research, Holland: Reidel Publishing.

FLOWERS, B. (1976) Nuclear Power and the Environment, Sixth Report, Royal Commission on En-

vironmental Pollution, London: HMSO.
FLYNN, C. B. and CHALMERS, J. A. (1980) The Social and Economic Effects of the Accident at Three Mile Island, NUREG/CR-1215, Tempe, Arizona: Mountain West Research.
GILLETTE, R. (1972) 'Nuclear safety: the roots of dissent', Science, 177, 771-776.
GEORGE, B. V. and HILSLEY, D. C. (1982) 'Britain's approach to the PWR stresses safety and reliability', Nuclear Engineering International, 21, 34-46.
GORDON, R. L. (1984) 'Reforming regulation of electric utilities in the USA: priorities for the 1980s', Energy Policy 12, 146-156.
HCSCE (House of Commons Select Committee on Energy) (1981), Report on the Nuclear Power Programme, London: HMSO.
HLSCEC (House of Lords Select Committee on the European Communities) (1984) European Community Energy Strategy and Objectives, London: HMSO.
HOWELLS, G. D. and GAMMON, K. M. (1984) 'Role of research in meeting environmental assessment needs for power station siting', in R. D. and T. M. ROBERTS, (eds.) Planning and Ecology, London: Chapman and Hall.
HSE (Health and Safety Executive) (1982) Emergency Plans for Civil Nuclear Installations, London: HMSO.
JAWORSKI, E. and WARD, R. M. (1984) 'Time cost/relief in nuclear power plant construction: public policy lessons from Enrico Fermi 2 in Michigan?' Paper presented at AAG Conference, April, Washington DC.
KASPERSON, R. E., BERK, G., PIJAWKA, D., SHARAF, A. B. and WOOD, J. (1980) 'Public opposition to nuclear energy: retrospect and prospect', Science, Technology and Human Values, 5, 11-23.
KATZ, J. E. (1984) 'US energy policy: impact of the Reagan Administration', Energy Policy 12, 135-145.
KEMENY, J. G. (1980) Report of the President's Commission on the Accident at Three Mile Island, New York: Pergamon.
LARKINS, J. T. and CUNNINGHAM, M. A. (1983) Nuclear Power Plant Severe Accident Plan, NUREG 0660, Washington DC: Nuclear Regulatory Commission.
LOWE, P. and GOYDER, J. (1983) Environmental Groups in Politics, London: George Allen and Unwin.
MACRORY, R. (ed) (1982) Commercial Nuclear Power:

Legal and Constitutional Issues, London: Imperial College.
MARLEY, W. G. and FRY, T. M. (1955) 'Radiological hazards from an escape of fission products and implications in power reactor locations', in Proceedings of the International Conference on Peaceful Uses of Atomic Energy, New York: United Nations, 102-105.
MONOPOLIES AND MERGERS COMMISSION (1981) Central Electricty Generating Board: a Report on the Operation by the Board of its System for the Generation and Supply of Electricity in Bulk, London: HMSO.
NRC (Nuclear Regulatory Commission) (1979) Report of the Siting Policy Task Force, NUREG 0625, Washington DC: Nuclear Regulatory Commission.
NRC (Nuclear Regulatory Commission) (1980) NRC Action Plan Developed as a Result of the Three Mile Island Accident, NUREG 0660, Washington DC: Nuclear Regulatory Commission.
NRC (Nuclear Regulatory Commission) (1981) Scoping Summary Report: Environmental Impact Statement on the Siting of Nuclear Power Plant, Washington DC: Nuclear Regulatory Commission.
OPENSHAW, S. (1982a) 'The siting of nuclear power stations and public safety in the UK', Regional Studies, 16, 183-198.
OPENSHAW, S. (1982b) 'The geography of reactor siting policies in the UK', Transactions of the Institute of British Geographers, New Series, 7, 150-162.
OPENSHAW, S. (1986) Nuclear Power: Siting and Safety, London: Routledge and Kegan Paul.
OUTER CIRCLE POLICY UNIT (1979) The Big Inquiry, London: Outer Circle Policy Unit.
OWENS, S. E., HILLS, P. R. and COPE, D. R. (1983) Energy Policy and Land Use Planning, London: SSRC.
PALLADINO, N. J. (1981) Remarks to Atomic Industrial Forum Annual Conference, December 1st.
PALLADINO, N. J. (1984) Remarks made at AIF/ANS/ENS Conference, Washington DC, November 14th.
PEARCE, D., EDWARDS, L. and BEURET, G. (1979) Decision Making for Energy Futures, London: MacMillan.
PRINGLE, P. and SPIGELMAN, J. (1983) The Nuclear Barons, London: Sphere Books.
PURDUE, M., KEMP, R. and O'RIORDAN, T. (1984) 'The context and conduct of the Sizewell B Inquiry', Energy Policy, 12, 276-282.
QUIRK, J. and TERASAWA, K. (1981) 'Nuclear regula-

tion; an historical perspective', Natural Resources Journal, 21, 833-855.
ROLPH, E. S. (1979) Nuclear Power and the Public Safety, Toronto: Lexington.
ROWAN-ROBINSON, J. and EDWARDS, J. (1981) 'The special inquiry', Town and Country Planning, 50, 79-80.
STARR, C. and BRAUN, C. (1984) 'US nuclear power performance', Energy Policy, 12, 253-258.
UCS (Union of Concerned Scientists) (1984) Nuclear Power Plants in the United States: Current Status and Statistical History, Boston: Union of Concerned Scientists.
USAEC (US Atomic Energy Commission) (1962) Reactor Siting Criteria 10CFR Part 100, Washington DC: US Atomic Energy Commission.
USDOE (US Department of Energy) (1984) Nuclear Power Program Information and Data, Washington DC: US Department of Energy.
US OFFICE OF TECHNOLOGY ASSESSMENT (1984) Nuclear Power in an Age of Uncertainty, OTA-E-216, Washington DC: US Office of Technology Assessment.
WALSKE, C. (1984) 'A perspective on nuclear power in the USA', Atom, 227, 7-9.
WYNNE, B. (1982) Rationality and Ritual: The Windscale Inquiry and Nuclear Decisions in Britain, British Society for the History of Science; Chalfont St. Giles, Bucks.
ZEIGLER, D. J. and JOHNSON, J. H. (1984) 'Evacuation behaviour in response to nuclear power plant accidents', Professional Geographer, 36, 207-215.

PART 2:

NUCLEAR WASTE: THE ACHILLES HEEL

Chapter Six

OUT OF SIGHT, OUT OF MIND: THE POLITICS OF NUCLEAR WASTE IN THE UNITED KINGDOM

Andrew Blowers and David Lowry

Power and Policy

It is now widely recognised, both by promoters and opposers of nuclear power alike, that the problem of managing radioactive wastes is the critical 'gap' in plans to expand the British civil nuclear programme. The problem in purely physical terms is becoming critical because of the ever expanding volumes of nuclear wastes that have been created since the inception of the British atomic energy project in the 1950s. The early wastes resulted from research and development of the British nuclear weapons programme; but the increase in volume is due to wastes created in the MAGNOX and Advanced Gas-cooled (AGR) reactors of the nuclear electricity programme, along with detritus from the research installations of the United Kingdom Atomic Energy Authority (UKAEA) and, especially, those wastes from reprocessing operations at the British Nuclear Fuels plc (BNFL) Windscale plant at Sellafield, which accounts for four-fifths of the low level waste (LLW) and 40 per cent of intermediate level waste (ILW). On top of this there are undisclosed amounts of military waste. It is expected that the volume of wastes will further increase after the year 2000 as the present government supports a go-ahead for new reactor programmes based on Pressurised Water Reactors (PWRs), and, more distantly, on Fast Breeder Reactors (FBRs). Additionally, as the older nuclear stations are shut down, the process of decommissioning them will result in a further substantial volume of radioactive wastes which must be dealt with. The legacy of 35 years or so of nuclear development in Britain has become a technical problem of burgeoning difficulty. The nuclear waste disposal problem has become the Achilles Heel of the nuclear industry.

The power to determine nuclear waste management strategies has been exercised within an exclusive, interconnected set of relationships between government and the nuclear industry. For example, the Radioactive Waste Management Advisory Committee (RWMAC), which advises government on waste management, is made up predominately of people with relevant backgrounds of expertise and it includes members from the nuclear industry, the Central Electricity Generating Board (CEGB) and the appropriate trades unions. There is only one environmental scientist on the committee which, in one critic's view, is 'packed with representatives of the nuclear industry' (FoE, 1984, p. 74). Another independent advisory body, the Advisory Committee on the Safety of Nuclear Installations (ACSNI), is advised by assessors drawn from the nuclear industry. The National Radiological Protection Board (NRPB), established in 1970, which provides advice on radiological hazards and which will advise on the authorisation of any future waste disposal facilities, has received commissioned research from the body responsible for implementing waste disposal policy. This body, the Nuclear Industry Radioactive Waste Executive (NIREX), set up in 1982, is composed mainly of those organisations responsible for the creation of radioactive waste. Three of the NIREX directorate are also members of RWMAC. Although NIREX is not a Government body it is closely identified with the policies for nuclear expansion currently proposed by the Department of Energy. This department controls the activities of the CEGB, the South of Scotland Generating Board (SSEB) and BNFL who are the major producers of radioactive wastes and are also members of NIREX.

The Department of the Environment (DoE) is the Government Department with prime responsibility for ensuring safety. It is responsible for nuclear waste site authorisation, for licensing disposal and for determining planning applications for specific sites. While this Department is anxious to maintain its distance from the nuclear industry it is unlikely that it will expose itself to public conflict with the Department of Energy. Rather more likely is that informal negotiations will be conducted through the government machine. Ultimate responsibility for waste management policy rests with the Cabinet, at which the Ministers responsible for nuclear policy and nuclear waste disposal must settle their differences.

The nuclear industry possesses decisive advantages over its environmental and planning critics in the decision making process, through its membership of or privileged access to Government bodies. This enables positions to be reached on such crucial subjects as radiological risk assessment, waste management strategy, nuclear energy and reprocessing policy before exposure to public debate. The Government assumes that nuclear expansion is justified and so for the present it remains a non-issue. Similarly, debate on alternative strategies for waste management may be forestalled. In this way the nuclear industry can determine appropriate policies without the necessity of providing a full justification, either for nuclear strategy in general or for specific solutions to the waste disposal problem. In the nuclear field predetermined ends condition the means.

This concentration of power was able to control the development policy for a long period. But the nuclear industry has become more vulnerable to its critics who have deployed three lines of attack on the waste management issue. First, they maintain there is no waste management strategy, merely an incremental approach leading to a predetermined solution. Second, they argue that in the absence of a strategy there is no case for the expansion of either reprocessing or producing electricity from nuclear reactors. And, third, they recognise that existing and future levels of waste constitute a problem but that it should not necessarily be resolved by imposing the risk of radiation damage on communities where there are not existing nuclear installations. Even so, there appears to be no identifiable policy formulated by the critics for the safe management of nuclear waste arising from hospitals and radiological medical research establishments. In general the nuclear industry's opponents regard the proposals for nuclear waste management put forward by the Government and NIREX as a means of ensuring relatively cheap and swift solutions which will overcome a stumbling block in the way of the future of the nuclear industry.

This chapter examines the proposition that waste disposal strategy in the UK is a political rather than a 'rational' or 'scientific' process. It shows how strategy has evolved in response to the changing political environment, leading to the proposals for land burial of radioactive wastes. It explores the political controversy over this policy, with specific reference to the case of one of the proposed sites. The future implications of the contemporary conflict are discussed in the concluding sections.

1. THE HISTORICAL POLITICS OF NUCLEAR WASTE, 1949-84

The early years

As early as 1952 James Conant, then President of the American Chemical Society, asserted that nuclear energy would founder because the problem of radioactive waste disposal was insoluble. It is not surprising that a man of Conant's eminence—a former President of Harvard University and a member of the wartime US National Defence Research Committee that was intimately involved in the Manhattan Atomic Bomb Project - should make such a sombre and prophetic assessment, as he had direct access to the key atomic researchers of the era. Conant was one of a minute band of nuclear pioneers who attempted to see beyond the atomic euphoria that characterised the 1950s. Another sceptic was Professor George L. Weil who wrote in 1955:

> The beneficial prospects associated with the development of nuclear energy have been widely publicised. On the other hand, discussions of the unpleasant aspects have been limited almost exclusively to the technical meetings and publications (Weil, 1955).

It was Weil who extracted the first fuel rod from the first atomic reactor in Chicago, December, 1942. The most characteristic appraisal of the atomic future was well expressed by Glenn T. Seaborg, who discovered plutonium and was Chairman of the US Atomic Energy Commission from 1961 to 1971. Seaborg saw the future of civilisation in the hands of the nuclear scientists - 'An elite team that would go about building a new world through nuclear technology'. It was a vision of atomic powered plenty: atomic energy was the magician's potion that could free industrial society permanently from all practical bonds. Deserts could be made to bloom, sea water made drinkable, mountains moved, rivers diverted, automobiles would run for ever without changing fuel. No problem would remain unsurmountable in the Atomic Age. Amidst such a vision, 'Chemists and engineers were not interested in dealing with waste. It was not glamorous, there were no careers, it was messy, nobody got brownie points for caring about nuclear waste' so recalled Caroll L. Wilson, who was General Manager of the US Atomic Energy Commission from 1947-1951 (Wilson, 1979).

The official historian of the British Atomic Energy Project, Professor Margaret Gowing, records that similar sentiments prevailed in Britain (Gowing, 1964, 1974 a and b). In the first volume of her history covering 1939-1945, there is no mention at all of radioactive waste (Gowing, 1964). In her discussion of the second period, 1945-1952, reviewing all the documentation - even that kept secret after the '30 year rule' - she wrote:

> Atomic energy risks were treated as a technical not a political or social matter, and were firmly in the hands of experts, with politicians showing little interest except in the Radioactive Substances Bill of 1948 (Gowing, 1974, p. 92).

There are only two other references in the three volumes comprising this study. In other studies, such as those by Jay (1952, 1955, 1961), the Atomic Energy Research Establishment's historian, nuclear waste management barely rates a mention. In these early years, 'To be cautious was to be unpatriotic' (Pringle and Spigelman, 1983). But, there is some evidence that nuclear waste was already a serious problem in these years. The first known accident involving nuclear waste occurred at a waste dump at Kyshtym near Chelyabinsk in the Soviet Union in 1958 or 1959. It received scant publicity at the time; a three paragraph report in the _Evening News_, London, of 18 March 1959 (reporting an article in the Viennese newspaper, _Die Presse_). This was the only mention until the Soviet scientist Zhores Medvedev published a fuller account many years later (Medvedev, 1976, 1977, 1979). The then UKAEA Chairman, Sir John Hill, described Medvedev's account as 'science fiction' and 'rubbish' indicating how the world's nuclear establishment (for it is important to remember that atomic energy in its early years was a truly global collaboration) kept the accident quiet for fear of alerting people to the darker side of nuclear power (Trabalka _et al_., 1979).

At about the time of this accident Britain was establishing her first policy guidelines for the management of nuclear wastes in a White Paper (HMSO, 1959). This was followed by the Radioactive Substances Act of 1960 which came into effect in 1963. Other relevant legislation was the Nuclear Installations Act of 1965 and the Radiological Protection Act of 1970 which established the NRPB. But it was fully twenty years before the guidelines in the 1959 White Paper were re-analysed by an 'expert group' drawn from Government Departments and organisations concerned with radioactive waste management (HMSO, 1979). International concern was also slow to develop. In 1957 the Organisation for European Economic Cooperation and Development (OEECD) was set up to include Britain, and it established a European Nuclear Energy Agency to foster the orderly development of nuclear energy. The Agency set up committees on radiation protection and safety, but only in 1975 was it felt sufficiently pressing to establish a Radioactive Waste Management Committee (RWMC). This body facilitated information sharing among member states and sponsored research (the equivalent UK body was first established in 1976).

The UK was slow to respond to international developments. In 1957 the EEC had set up EURATOM, its nuclear agency, but the UK did not fall within its auspices until signing the Treaty of Rome in 1973. The UK had, however, signed a United Nations Declaration in 1958 which stated:

> Every state shall take measures to prevent pollution of the seas from the dumping of radioactive waste, taking into account any standards and regulations which may be formulated by the competent international organisation (US Convention on the High Seas, Geneva, Article 25 (1)).

Despite this, for a period of thirteen years, 1950-63, the British Government had been dumping radioactive wastes into the sea, ten miles north of Alderney in the Channel Islands - a site which had also been used to dump surplus and degraded high explosives from World War II. This information came to light only in June 1984. The bulk of the waste must have come from the nuclear weapons programme though some could conceivably have originated in the early civil research project. During the 1950s, Britain was also flouting international opinion on protecting the maritime environment from radiological contamination through the discharge of radioactive materials from Windscale. J. H. Dunster, then a scientist with the UKAEA's Industrial Group told an international conference that the levels were high enough 'to obtain detectable activity levels in samples of fish, seaweed and shore sand' (Dunster, 1958, p. 391). Dunster even added that in 1956 'the rate of discharge of radioactivity was deliberately increased partly to dispose of unwanted wastes, but principally to yield better experimental data' (ibid., p. 398). In effect a Government agency, the UKAEA, was using the Irish Sea as a large-scale radiological experiment. Mr Dunster is now Director of the NRPB.

International opinion led to strengthened demands for environmental protection. The UN Conference on the Human Environment in Stockholm in 1972, called for control, as did the so-called London Dumping Convention (LDC), meeting for the first time that year. The British Government ratified the LDC in 1975, yet in 1976 it released plans to increase, by thirty to forty times, the amount dumped in the sea by the 1990s. From 1975 to 1981 this amount dumped doubled. But the political climate was changing and the Government was forced to change its waste disposal plans. In 1983, twenty years after the first unregulated sea dumping programme ended, the National Union of Seamen, aided by the environmental group Greenpeace (UK), brought to an unceremonious end the official UK dumping programme. The success of this campaign created the conditions under which subsequent waste disposal policy developed.

Sea dumping: an option foreclosed

Each year the global environmental organisation Greenpeace chose to make the meeting of the LDC in London a focus for highlighting its opposition to the continued British annual dump of low and intermediate level waste (LLW & ILW) in the North Atlantic. Thus in mid-February 1983, demonstrators clad in radiation contamination-proof suits lobbied the LDC at the International Maritime Consultative Organisation in London, opposite the Houses of Parliament. At the meeting the Spanish delegation spoke strongly against the British dumping policy. The next day (16 February) an editorial in *The Times* pronounced:

> There are no local constituents to complain

> when waste is dumped in mid-ocean. But it will not be possible either technically or politically to go on using the sea indefinitely.

On February 22 the Agriculture Minister, Alec Buchanan-Smith, told Parliament why Government representatives to the LDC had voted against a halt to sea dumping:

> We did so because the relevant resolution involved action that was against the policy of Her Majesty's Government and against the spirit of the Convention. In particular, it called on member states to act in advance of receiving authoritative scientific advice: to do so would be contrary to the provisions of Article XV (2) of the convention.

This represents a remarkable piece of sophistry; arguing that sea dumping should continue unless and until scientific advice demonstrated its unsuitability. The Government's rationalisation of its stance was clearly illustrated in a series of statements in its 1979 review of waste disposal policy (HMSO, 1979) from which we selected the following examples with our own comments in parenthesis:

1. We recommend that the UK should continue to seek to develop realistic international standards for disposal of low and intermediate level waste at sea. We believe that there can be quantitative justification for an increased sea dumping programme and we recommend urgent research to build up a body of knowledge which will demonstrate this. We also consider that the UK should meanwhile begin, modestly, but continuously, to increase sea disposals (p. 118 para. 7.23)

(i.e. before the research findings were completed).

2. We recommend that ... to permit realistic plans to be made for waste management of plutonium contaminated materials, strategy for continuation and expansion of the sea dumping programme should be agreed and implemented as soon as possible. (p. 102 para. 6.22). There is no evidence of irradiation of man following these disposals at sea ... we strongly recommend that the UK should continue to use this

disposal route (p. 79 para. 5.19).

(This assessment ignores the potential for wider radiological contamination of the ocean bed or the ecology of the ocean and merely concentrates on its human impact.)

3. The UK also should begin gradually to increase the sea disposals in accordance with the London Convention and supported by appropriate environmental investigation (p. 80 para. 5.21).

(This not only contravenes the essence of the 1976 Sea Dumping Convention, but in a completely unscientific manner suggests that the proposed investigation even before it has taken place, will support the planned increase of sea dumping. This type of argument is also seen in the following extract.)

4. In our view, sea disposal will probably prove to be the best option in terms of cost and radiological protection; we anticipate that the detailed studies leading to a decision for each type of waste will confirm this. As we have already said, this method of disposal is underused, but the international climate is such that it will be necessary to justify any substantial increase in sea disposal with more scientific evidence than is currently available. We recommend that the appropriate research and measurements to build up the necessary body of knowledge to do this should be pursued urgently. We recommend that the present level of disposing of intermediate level wastes at sea should meanwhile be modestly but continuously increased above the present levels (p. 109 para. 6.32).

(Such attempts to pervert scientific analysis in order to provide justifications for policy decisions already taken are unfortunately not unique. Consider the following:)

5. It was suggested that the Public Health Service come up with its views as to what levels would correspond to enough of a health risk to justify diversion of resources in order to provide protection. If any reasonable agreement on this subject can be reached among the agencies, the basic approach to the report would be to start with a simple, straightfor-

> ward statement of conclusions. It would then be a straightforward matter to select the key scientific consultants whose opinions should be sought in order to substantiate the validity of the conclusions or recommend appropriate modifications.

(This statement comes from a study entitled 'Status Report of the Current Activities of the Federal Radiation Council Working Group' sponsored by the LDA Office of Radiation Programs in the United States. It is a clear illustration of the premature validation of policy which has characterised nuclear decision-making. In short the political ends justify the scientific means.)

However, the opposition to sea dumping plans was not to be undermined or to wither away. In mid-June 1983 it was revealed that three leading transport unions (Transport and General Workers (TGWU), Association of Locomotive Engineers and Firemen (ASLEF), and National Union of Seamen (NUS), would boycott the planned sea dump of nuclear waste due on the 11 July. The unions stuck to their decision despite a direct appeal to the NUS General Secretary, Jim Slater, by the sea-dump ship owners. Union Jacks were burned in Galicia by Spanish demonstrators in support of the Unions' action and other Spanish protests followed. On 15 July a group of anti-nuclear demonstrators invaded the Sizewell 'B' Inquiry (at the time in temporary residence in London) as the Department of Environment witness, George Wedd, was giving evidence on radioactive waste management.

Despite the furore that brought a continuous correspondence in the serious press it seemed that the Government had not abandoned hope of resurrecting the sea dump for 1983. Indeed, as part of a series of exchanges in the correspondence columns of The Times Dr. L. E. J. Roberts of the UKAEA and NIREX reiterated that sea dumping, adopted by successive UK governments, 'was soundly based on technical scientific observation and analysis'. By the end of August, however, it was being reported that all further plans for nuclear waste dumping at sea had been abandoned by the British Government. The Observer commented (28 August) that it was 'a remarkable victory for Greenpeace'. Within days Greenpeace added a further twist to the tale by releasing leaked confidential documents that showed that the Ministry of Defence (MOD) planned to repackage military nuclear wastes and dispose of them in the Atlantic. (Guardian 1, 2 September, New Statesman, 2 September). When Jim Slater took a resolution to halt dumping to the Trades Union Congress (TUC) in Blackpool, it was endorsed by 7,150,000 to 2,764,000 votes. Thereafter, despite further reported attempts by the Government to resurrect sea dumping plans, the option to use the oceans to dispose of radioactive waste was foreclosed.

In autumn 1983 the nuclear industry's reputation suffered its greatest damage for two decades (since the 1957 fire at Windscale) with the radioactive leaks from Sellafield (see Chapter 10) and the implied connection between Sellafield and the high incidence of childhood leukaemias in the area (see Chapter 11). The impact of these events in transforming the political environment has been described in Chapter 1. It was, therefore, no great surprise that in early December the Government acceded to demands for a joint Government-TUC Committee of investigation into sea dumping and its potential dangers. By the end of 1983 the long-term campaign by Greenpeace to end sea dumping was vindicated. A year later when the Government-TUC Committee under Professor Fred Holliday reported (HMSO, 1984) it recommended that the moratorium on sea dumping should remain in force and the Government acceded. The sea dumping campaign had combined international and national opposition to undermine the power of the government-nuclear industry complex to pursue its desired waste disposal policy. In doing so it had narrowed the available waste disposal options. During this period the political climate had begun to shift against the nuclear industry, resulting in further reverses for Government policy as the following section shows.

Waste disposal - options deferred

> Somewhere, somehow the dream of the 1950s and 1960s that nuclear power would become the predominate source of electricity by the end of the century was shattered (Green, 1982).

In the late 1960s the benefits of nuclear energy began to be questioned (Falk, 1982). In the West, at least, it was a period of general questioning by the youth culture of the wisdom of their elders. Nuclear energy, previously unassailable, came under scrutiny along with a panoply of previously sacred cows. It was a painful experience for the nuclear industry, which hitherto had enjoyed almost universal acclaim, especially as nuclear electricity had meant 'Atoms for Peace' and hence not 'Atoms for War' - to the general public at least.

In the general scientific press of the early 1970s the lack of a sustainable strategy for nuclear waste management was criticised. This contrasted strongly with the purely technocratic and uncritical expositions of earlier years. In the New Scientist (23 August 1973, p. 456) appeared a critique of the large Hanford nuclear complex in Washington State, USA, where plutonium contamination of a burial ground and adjacent river was found to be extensive. That autumn there was a major report called 'The Nuclear Fuel Cycle' by the Boston-based Union of Concerned Scientists. It highlighted the nuclear waste disposal problem that had barely been discussed before outside the literature of the nuclear industry.

In the UK, too, a group of 28 scientists and policy makers explained the significance of the radioactive waste management problem:

> The debate, last year, about the choice of reactor systems in the development of a nuclear power programme obscured the problems posed by the production of dangerous long lived radioactive wastes which have to be kept away from the living environment for hundreds, and some for many thousands of years (letter to the Guardian, 7 January 1974).

This statement encapsulated the problem that thenceforth was to haunt the nuclear industry; but now it was to be a public haunting - and a public conflict. Articles, correspondence and, later, television programmes and radio debates followed. In October 1975 the Daily Mirror ran a page 1 article that began: 'A storm of protest is growing over secret plans to import huge quantities of nuclear waste into Britain' - (Daily Mirror, 21 October 1975). That this appeared in a newspaper that supported the Labour Party - whilst the Labour Party was in office - sent a frisson up the spines of nuclear executives in Britain; a 'frisson' that would spread abroad to the industry's foreign customers. (In fact, the issue had been headlined five months earlier on the front page of a Friends of the Earth tabloid, Nuclear Times, but it had made little impact since Friends of the Earth had not then been accepted as a major source of criticism of the industry.) In one sense it can be seen as the first major salvo in the debate that preceded the Windscale Inquiry into reprocessing - held from June to November 1977 - a debate that has intensified in the 1980s as Chapter 1 has shown.

The nuclear waste debate began to encompass the variety of disposal options: sea dumping; deep and shallow land burial. It also involved critics from the respected and respectable establishment as well as environmental groups. In particular the Sixth Report of the Royal Commission on Environmental Pollution (the Flowers Report), published in 1976, contained the following oft-cited paragraph:

> There should be no commitment to a large programme of nuclear fission power until it has been demonstrated beyond reasonable doubt that a method exists to ensure the safe containment of long lived highly radioactive waste for the indefinite future (HMSO, 1976).

Opponents of nuclear power took this as support for their case that nuclear power should be stopped. The nuclear industry, picking up on the words 'demonstrated beyond reasonable doubt', took it to mean a green light for the practical exploration of disposal options. Thus, in 1976, the UKAEA initiated a long-range programme to ascertain what potential sites existed in the UK that were suitable (on geological grounds) for the disposal of high level nuclear wastes that were building up at the Windscale site in Cumbria. Additionally the Department of the Environment set up a new Nuclear Waste Management Division in March 1978 - in part due to the pressure from public opposition.

The UKAEA's attempts to carry out this programme brought it into confrontation with local communites in Mullwarcher in Ayrshire (SCRAM, 1980); in mid-Wales (Arnott, 1982) and in the Cheviot Hills in Northumberland (TUSIU, 1980). On each occasion the nuclear industry's plans were thwarted. The opposition combined radical environmentalists, working land-owners, farmers and Conservative conservationists. This broad alliance was supported by detailed scientific advice from some of the few nuclear waste disposal experts outside the nuclear industry or the government-sponsored regulatory bodies. The Government's proposals were finally abandoned in December 1981, when the Secretary of State for the Environment announced that the drilling plans for the prospective HLW sites were to be wound up despite protests by the Institute of Geological Sciences.

The opposition had scored some notable successes in preventing further sea dumping and in halting any progress on HLW disposal sites. In 1983 a third battle over nuclear waste began with the announcement of potential sites for the land burial of ILW and LLW at Billingham in the industrial North East of England and at Elstow in the English lowlands in Bedfordshire. In the second part of this chapter we examine the proposals for dealing with HLW and ILW in the changing political context of the 1980s.

2. THE CONTEMPORARY POLITICS OF NUCLEAR WASTE

Land disposal - the Government 'strategy'

By 1983 sea disposal of nuclear wastes had been, perforce, abandoned and the disposal of HLW had been deferred for at least fifty years. Nevertheless, the Secretary of State for the Environment claimed that the effective disposal of nuclear wastes, 'in ways which have been shown to be safe, is well within the scope of modern technology' (Hansard, 25 October 1983). The remaining problem, ILW, had been solved since, 'The Government believe that land-based disposal of intermediate wastes is the safest and best method, provided that a site can be found with a sufficient geological certainty and stability which will remain safe for the necessary period of time' (Ibid.).

The origins of this strategy can be perceived in a number of Government statements. In its response to the Royal Commission's Sixth Report, the Government had accepted the need for a strategy based on minimising the creation of wastes; ensuring that waste management was dealt with before a large nuclear programme was undertaken; recognising concern for environmental considerations; and undertaking adequate research and development on disposal methods. In particular the Royal Commission advocated disposing of existing accumulated wastes 'in appropriate ways, at appropriate times and in appropriate places' (HMSO, 1976, para. 14). Within six years of the Royal Commission's Report the Government produced a White Paper which claimed that, 'waste management is not a barrier to the further development of nuclear power as foreseen' (HMSO, 1982, para. 24). In that intervening period (1976-1982) the 'preparation of an overall long-term strategy' (para. 22) had commenced. Two advisory committees, the Radioactive Waste Management Advisory Committee (RWMAC) and the Advisory Committee on the Safety of Nuclear Installations (ACSNI), had been established and research expenditure had increased. But instead of an independent body dealing with waste management, as envisaged by the Royal Commission, the Government announced the formation of the Nuclear Industry Radioactive Waste Executive (NIREX) with a membership drawn from the nuclear industry and the generating boards.

Two important objectives underlay the White Paper. These were, first, 'That any practice giving rise to radiation exposure must be justified in terms of its overall net benefit to society'; and, second, 'that all exposures should be reduced to levels which are as low as reasonably achievable, economic and social factors being taken into account' (para. 11, the so-called ALARA [as low as reasonably achievable] principle).

The first objective is rarely questioned, being regarded as implicit in the Government's pursuit of a nuclear energy policy. Indeed, the CEGB, is quite candid on this issue as it affects waste disposal: 'No "justification" is necessary for waste disposal because the generation of electricity by nuclear power is considered a net benefit.' (CEGB 1982a, p. 39). Certainly the existence of nuclear power stations and their accumulating wastes had created a problem which required a solution. But the justification of future nuclear energy expansion and reprocessing is highly contested on grounds of cost, needs, safety, environmental impact and waste disposal. It is clear that the future development of nuclear facilities ought to be justified in terms of an overall social benefit sufficient to outweigh the specific social costs imposed upon populations surrounding the facilities.

The second objective, the ALARA principle, is site specific. This principle is deeply enshrined in the British approach to pollution control, whereby Government inspectorates seek the 'best practicable means' of pollution reduction by weighing up the relative costs and benefits of individual cases (see Ashby and Anderson, 1981; Rhodes, 1981; Blowers, 1984; Hawkins, 1984). This approach is distinctively different from that usually adopted in other countries where standards are set which must be achieved using the best available technology (BAT) regardless of economic and social considerations. It can be argued that radiation poses a different dimension of hazard from any other known form of pollutant. Some forms of radioactive materials are dangerous for hundreds, even tens of thousands, of years. It is true that, unlike some toxic materials, they decay over time. They are different in another respect: 'Most chemical wastes of a non-radioactive kind are only dangerous if ingested or brought into contact with the skin, whereas we know that some kinds of radioactive waste can injure living tissues by proximity' (Sizewell 'B' Inquiry Proceedings Day 100, 9-10). At present the Government is proposing the less stringent interpretation of ALARA but, as we shall see, the NIREX proposals fail even on this criterion.

The Government's strategy appears to consist of the following elements. HLW will remain at Sellafield in tanks until vitrification is introduced in the 1990s while disposal options are identified so that future generations 'will be better placed than we are to make the eventual choice' (HMSO, 1982, para. 5). LLW will continue to be disposed of at Drigg in Cumbria and at a future inland disposal site, and possibly by resuming sea disposal. In the case of ILW, hitherto stored at various sites: 'There is no technical obstacle to the development of the disposal facilities required' (Ibid. para. 15). The future expansion of nuclear power and of reprocessing, and the looming problem of decommissioning the ageing MAGNOX plants, would therefore not be hindered by waste disposal problems. In a remarkably short time the Government seemed to have moved from a position of having virtually no strategy to a situation where it felt able to claim that a strategy had been sufficiently evolved to a point where: 'The problems posed by radioactive wastes are of a manageable nature' (DoE, 1984a, para. 50). Not only had the Government and the nuclear industry successfully rationalised the available options to one of land disposal of LLW and ILW, they had gone still further by reducing the available sites for disposal to a shortlist of two. Both the general strategy and the specific proposals were flawed by inconsistencies and contradictions which are explored in the following sections.

Figure 6.1

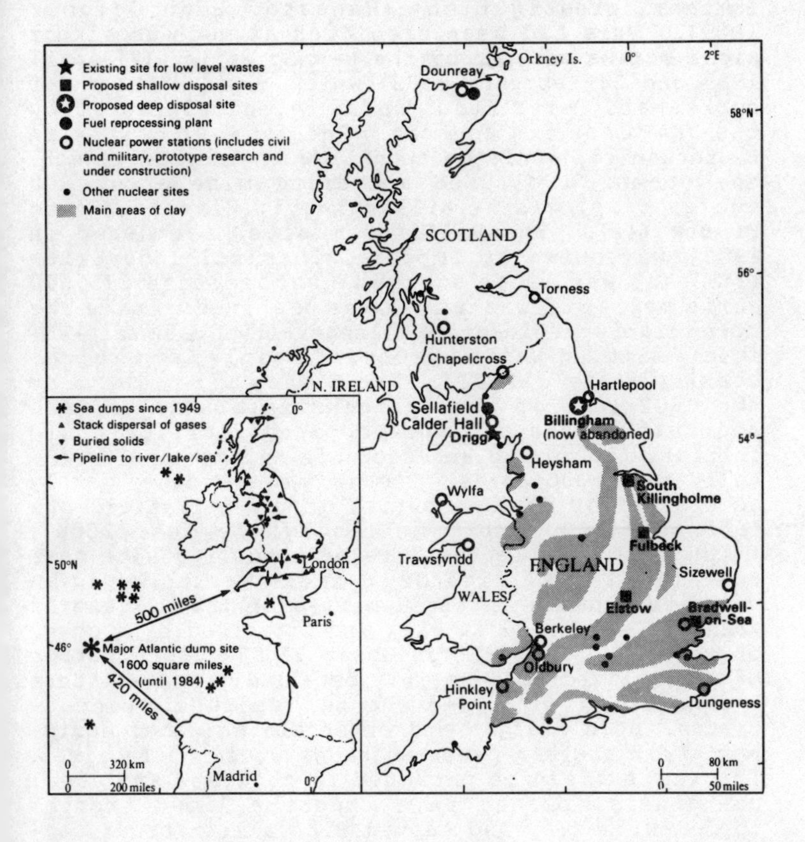
Existing site for low level wastes
Proposed shallow disposal sites
Proposed deep disposal site
Fuel reprocessing plant
Nuclear power stations (includes civil and military, prototype research and under construction)
Other nuclear sites
Main areas of clay
Sea dumps since 1949
Stack dispersal of gases
Buried solids
Pipeline to river/lake/sea
Orkney Is.
Dounreay
SCOTLAND
Torness
Hunterston
Chapelcross
N. IRELAND
Sellafield
Calder Hall
Drigg
Hartlepool
Billingham
(now abandoned)
Heysham
South
Killingholme
Wylfa
Fulbeck
Trawsfynydd
ENGLAND
Sizewell
WALES
Elstow
Bradwell-on-Sea
Berkeley
Oldbury
Hinkley Point
Dungeness
London
Paris
Madrid
500 miles
420 miles
Major Atlantic dump site
1600 square miles
(until 1984)
50°N
46°
0°
2°E
58°N
56°
54°
320 km
200 miles
80 km
50 miles

Land disposal - the selection of sites

In October 1983 the Secretary of State for the Environment announced that NIREX had identified two sites as being sufficiently 'promising to justify further investigation' (Hansard, 25th October 1983). This had been predicted as much as six or eight months earlier by the New Scientist (21 April 1983 and 24 February 1983) which cited NIREX chief spokesman, Peter Curd, and a technical report by the UKAEA, as its sources. One site, at Elstow in Bedfordshire, would be a shallow engineered trench; the other, a disused anhydrite mine about 300 metres deep, was at Billingham in Cleveland (see Figure 6.1). The Billingham site, abandoned in 1971, was owned by Imperial Chemical Industries (ICI) and was large enough to accommodate 275,000 cubic metres of waste. It was destined to take the more highly radioactive, longer-lived alpha particle-emitting wastes produced mainly from reprocessing at Sellafield. The Elstow site, owned by the CEGB (one of the partners in NIREX), covers about 450 acres and the proposed facility would initially occupy 60 acres or so, and perhaps eventually about 300 acres. Elstow was envisaged partly as a site for shallow burial of non-Sellafield LLW (about 500,000 cubic metres by the year 2000), which would enable the life of the Drigg site near Sellafield to be extended. These wastes would be deposited to within three metres of the surface of trenches excavated in clay and protected by a cover of topsoil. In addition about 25,000 cubic metres of short-lived ILW - mainly beta and gamma emitters - consisting of ion-exchange residues, concentrates, pond sludges and other contaminated equipment from nuclear power stations would be received. The waste would be packaged in concrete and steel containers, placed in a concrete lined trench, topped with clay and capped with a reinforced concrete intrusion shield and further layers of clay, concrete and asphalt bitumen. The decision to select Elstow is difficult to comprehend as a rational process, taking into account the various criteria nominated both by Government and NIREX. In terms of safety, location, hydrogeological and land-use and environmental considerations it appears unlikely that it can be justified as a preferred site. These criteria are examined for Elstow in the following sections.

Safety

NIREX's prime objective is that sites for waste disposal should ensure that risks are as low as reasonably achievable. Consequently, greatest emphasis is given to hydrogeological factors. Indeed these are the only factors which were considered in any depth in the initial project statement. The overriding principle is that 'The facility should not give rise to any risk to future generations greater than would be acceptable to the current generation' (DoE, 1984b, 4. 1a). Risk is defined by the National Radiological Protection Board (NRPB) as 'a combination of the probability that a dose will be received and the probability that the dose will give rise to deleterious effects' (NRPB, 1984, p. 21). The DoE has stated that the future risk of death from radiation to the individual should be no greater than 1 in 100,000 in any year from all sources but that the target for an individual repository is 1 chance in 1 million from the repository itself.

The NRPB is the body responsible for advising on the radiological standards to be applied to waste management having regard to the recommended criteria of the International Commission on Radiological Protection (ICRP). However, it is worth noting that at the Sizewell 'B' Public Inquiry the lack of independence of NRPB and ICRP scientists from their colleagues working still within the nuclear industry was strongly criticised (Sizewell 'B' Inquiry, 1984a). So far the NRPB has engaged in generic studies using models based on a number of simplifying assumptions. In a 1982 assessment of the risks involved in disposing of PWR operating wastes in shallow land facilities, the NRPB calculated that for the first 200 years or so the greatest risk would be to those working on the site, through inhalation of dust and external radiation. This risk would reduce with radioactive decay and the risk through the migration of radionuclides into the food chain would increase for up to 800 years. This report stressed that its predictions were tentative and that a great deal more research would be necessary into the hydrogeology of specific sites, the behaviour of radionuclides in the geosphere and their transport through the biosphere. Although the report concluded that shallow burial 'could be a radiologically acceptable option' it 'should not be used as an input to decisions on the acceptability of any specific burial site or facility design' (NRPB, 1982, p. 21).

Sixteen months later, in a report by the same authors, similar qualifications are made but the conclusion strikes a more confident note; 'The results do indicate that shallow burial in appropriate facilities, of reasonable quantities of (LLW and short-lived ILW) waste considered in this report is likely to be acceptable from a radiological protection viewpoint' (NRPB, 1984, p. 29). While the use of the site after closure would need to be restricted for only about 100 years in the case of LLW to ensure a dose equivalent limit of 5mSv to a critical group (equivalent to an average of 1mSv), in the case of certain ILW wastes the same dose limit would require a very long period of site restriction, at least 500-600 years (note 1). It should be noted that the calculation of dose equivalent are sufficiently close to the limit to allow little margin for error. Quite aside from the various simplifying assumptions in the report, a number of other possible hazards, such as microbiological activity and structural degradation of the site, were not taken into account.

Despite the NRPB's qualifications, its reports are used to legitimate the intentions of what we may collectively call the nuclear industry. The CEGB, in presenting evidence at Sizewell, commented that the 1982 report 'reveals no reason why deep burial of ILW cannot be developed as a very safe disposal route' (CEGB, 1982b, p. 29). NIREX claimed, in 1983, that the 'use of existing technology combined with sound management techniques can lead to the disposal of radioactive wastes in a very safe manner' (NIREX, 1983, p. 3). By 1984 the DoE itself was proclaiming: 'there is no advantage from delay, and no requirement for lengthy research and development into methods of disposal' (DoE, 1984a, para. 21). There is little evidence in the NRPB reports which supports such sanguine conclusions.

Serious contradictions remain between the conclusions of the NRPB reports and the revealed intentions of the industry. The CEGB has claimed that 70 to 80 years after closure of a site 'the land will be suitable for unrestricted rural use' (CEGB, 1982a, p. 38). NIREX argues that 'The process of radioactive decay ensures that the radioactivity of these wastes is reduced to very low levels over periods of, at most, two or three hundred years' (1983, p. 15). Such statements fly in the face of the estimates produced by the NRPB, and signal that in its haste to achieve land disposal the nuclear industry is placing future generations at levels of risk which are higher than those acceptable to the present generation. The lack of research, the conflicting evidence, and the different interpretations of likely future risks combine to suggest that land burial has been adopted as a preferred option long before the case could be properly justified.

Alternative sites

Not only had land burial been adopted, but also only two sites were initially identified by NIREX. This appeared to contradict the principles for the protection of the human environment drafted by the DoE and published simultaneously with the announcement of the sites at Elstow and Billingham (DoE, 1983), and later revised after consultation, (DoE, 1984b and c).

These principles suggested that although it would not be necessary for a developer to show that a proposed site was the best choice among alternatives, it would be necessary 'To show that he has followed a rational procedure for site identification and that in selecting his preferred site, he has not ignored a better option of limiting radiological risks' (Ibid., 1984b, 5.3). The earlier draft had also added: 'or for satisfying town and country planning criteria and policies' (1983, 5.3). It was clear that any proposal would have to be justified not merely in terms of its safety but also in terms of a comparative assessment of alternative sites. Site selection would need to demonstrate rational, comprehensive and comparative procedures.

At various points the need for comparative assessment of sites was recognised by Government Ministers. The Under Secretary of State for the Environment made it clear to a delegation from Bedfordshire that NIREX would be required to list a number of alternative sites and to say why it considered Elstow to be the most suitable. During an adjournment debate in the House of Commons he was quite unequivocal on the point: 'NIREX will continue to evaluate possible disposal sites in addition to those two sites. Any further sites that NIREX identifies will be announced as well' (Hansard, 3 May 1984). Unless such information was provided a ministerial decision would not be forthcoming. Despite these requirements NIREX conspicuously failed to reveal those alternative sites which it had considered. In its Project Statement on Elstow NIREX referred to 'about 60 potentially suitable repository sites' (NIREX, 1983, p. 1) which had been refined down to a short list from which Elstow emerged 'as most worthy of detailed consideration' (p. 10). The precise criteria employed by NIREX for its particular selection of Elstow were unclear. In 1985 NIREX was required by the Secretary of State 'to select and announce as soon as possible at least two further sites in addition to Elstow' (Hansard, 24 January 1985).

It was not until February 1986 that the Secretary of State announced three further sites, in addition to Elstow, for 'a possible near-surface facility for shorter-lived radioactive wastes'. These were near Fulbeck in Lincolnshire, at Bradwell in Essex and at South Killingholme in South Humberside (Hansard, February 25th 1986) (see Figure 6.1). Two possibilities arise. One is that these alternatives will simply provide a legitimation process for Elstow, the favoured site. The other is that after a comparative assessment Elstow will not be chosen, in which case it has to be asked why Elstow was considered in the first place as the most suitable site.

Elstow - Location

In presenting the case for Elstow, NIREX stated that the site was 'conveniently located' on a suitable geological formation (NIREX, 1983, pp. 1-2). Yet suitable clay formations are widely distributed throughout the south and midlands of England (Figure 6.1). A CEGB study (1982a) concluded that there were three areas where suitable clay was well located relative to power stations; in Worcestershire, the Northamptonshire-Peterborough area (the one nearest Elstow), and the Thames estuary. If only one site was required then Worcestershire was the best, whereas the best combination of two sites was in Worcestershire and north of the Thames estuary. The site nearest Elstow did not therefore emerge as an optimal location.

Looking more closely at the distribution of waste, much depends on the assumptions made about future nuclear power stations and their location, the future of reprocessing, the volumes arising from decommissioning and the volume reduction of wastes likely to be achieved through compaction. About three-quarters of LLW emanates from the labour-intensive reprocessing operations at Sellafield and, if reprocessing continues, then a site in southern England would not be near the main source of future waste. However, the intention is to use Drigg for Sellafield waste and the additional site for other LLW. But, if compaction were used at the Drigg site volumes there could be reduced by two-thirds thus extending the life of the site. Similarly, if reprocessing eventually ceased there would be remaining capacity at Drigg. In both cases the need for another site for LLW would be lessened. About 80 per cent of ILW is the long-lived highly-active ILW stored at Sellafield and destined for a deep burial facility such as that proposed initially at Billingham. The distinction between categories of ILW is murky and liable to shift, but the estimates all point to a relatively low proportion of ILW for Elstow and raise doubts about the need for shallow land facilities for such wastes in addition to other canvassed options.

Elstow - local siting features

In selecting Elstow, NIREX claimed that 'local hydrogeological conditions are relatively well-known' (NIREX, 1983, p. 10). But, borehole evidence from nearby areas suggests that the clay here is laminated rather than massive, subject to small faults and overlain by callow up to 4 metres thick which contains the water table, and underlain by permeable strata. Water movement in the aquifers appears to be slow but there are major gaps in the information about the site. It appears that the Oxford Clay at Elstow is one of the thinnest beds in the area and that, moving in any direction from Bedfordshire, the Oxford Clay beds increase in thickness. Borehole measurements, revealed after the NIREX announcement, established that the depth of clay at the site varies from 15.2 metres to 20 metres, including the callow overburden, with between 11.6 and 15 metres of Oxford Clay itself. Yet the 1982 White Paper suggested that the depth of clay should be 20-30 metres and the Department of the Environment (DoE) took 'about 30 metres depth' as the upper limit when the sites were first announced (DoE, 1983). This figure was later omitted from the revised document (DoE, 1984c), presumably so that it did not conflict with NIREX, which had completely ignored the DoE's criterion by stating in reference to Elstow: 'At this location about 15 metres of Oxford Clay is present'. Thus NIREX has not only selected a clay bed thinner than that suggested by government but one which is, in places, actually thinner than its own initial estimates.

On the question of land use the DoE made it quite clear that 'a site should be selected where it is unlikely that future development of natural resources, or of the site will disturb the facility' (DoE, 1984b, para. 4.1c). Yet the Oxford Clay in the Bedford area is already excavated by the London Brick Company for brickmaking. Long before NIREX existed, the precise site of the proposed repository had been earmarked by London Brick and Bedfordshire County Council as an area for future clay extraction for brickmaking (Blowers, 1984). Consequently, when it became clear that the CEGB, the owner of the site, was prepared to offer it to NIREX, London Brick applied for permission to develop the site together with adjacent land already in its ownership. London Brick thereby established a viable alternative use for the site, a use that was of national importance, and which would be lost if the NIREX proposals went ahead.

NIREX also stated that in arriving at initial sites it assessed proximity to population and impact on the environment. The site is close to the area south of Bedford which is designated for future residential expansion, and 120,000 people already live within five miles of it. Since there are other sites with smaller local populations the risks from Elstow cannot be as low as reasonably achievable using the criterion of population density.

It is clear that the selection of Elstow could not be explained simply by recourse to rational planning criteria. The whole 'strategy' being pursued possessed the appearance of an exercise in selecting both a disposal method and specific sites on the basis of political, not rational calculation. Therefore, the political context of decision making must be examined.

The local political context

Both Billingham and Bedford appeared initially to have been perceived as politically soft options for waste disposal. The two sites seemed available, accessible and cheap. The Elstow Storage Depot was owned by the CEGB and the Billingham anhydrite mine was owned by ICI, who were apparently willing to allow it to be considered for a waste disposal facility. Both sites are in areas where there are already concentrations of polluting industries. Billingham is in the heart of a major petrochemical and steel making complex containing 14.5 per cent of the country's registrable hazardous locations or, as the local MP put it; 'one in seven of Britain's gigantic dustbins' (Hansard, 3 May 1984). The Elstow site is in the Marston Vale where already 700 hectares of clay have been dug up, leaving deep and derelict pits, and where the air is polluted by sulphur dioxide and fluoride from brickmaking. Significantly, NIREX claimed that a repository at Elstow would be unlikely to 'affect the environmental character of the area' and that 'there is already a significant concentration of heavy goods traffic in the area' (NIREX, 1983, p. 11). Presumably, then, already-deteriorated environments could be used as a justification for adding further disbenefits.

There are also obvious differences between the two areas. Teesside is now Britain's worst unemployment blackspot. In time of severe recession environmentalist causes are unlikely to flourish, and so it may have been thought that opposition to a waste dump would be relatively weak in such a location. In the case of Bedford, one of the local MPs, Sir Trevor Skeet, was well known to be a strong supporter of the nuclear industry. When the site was first announced he declared that some two years previously he had suspected that the Elstow site would be nominated (*Bedfordshire Times*, 10 Nov. 1983). During the debate after the announcement Sir Trevor Skeet fielded the Secretary of State a series of '"soft" questions designed to give the Minister an easy ride by inviting him to agree with an innocuous proposition' (*Bedfordshire, on Sunday*, 30 Oct. 1983). Chairing a public meeting in Bedford, he said; 'It is perfectly natural that you should have anxieties if you are not knowledgeable on the subject' (*Bedfordshire Times*, 27 Nov. 1983). He refused to declare his opposition to the plans arguing that he would keep an open mind and weigh up the evidence. When the other sites were announced in 1986 he declared: 'I intend to examine them with some care. I will look at them individually and would be very interested in the proximity to population' (*Bedfordshire*

Times, 27 Feb. 1986), believing that Elstow would be ruled out on grounds of the dense population nearby. Mr Nicholas Lyell, the MP for Mid-Bedfordshire, which includes the Elstow site, though a supporter of nuclear power, declared his opposition to the Elstow proposal from the outset.

If NIREX calculated that it had found sites where protest was likely to be muted, it had profoundly miscalculated. The hostility was immediate and virtually unanimous. In Billingham, where rumours that the mine would be chosen were circulating months before the announcement, local public and political opinion was orchestrated by a campaign, Billingham Against Nuclear Dumping (BAND). The campaign included the panoply of tools deployed by protesters - public meetings, vigils, letters to MPs, a delegation to Westminster, a mammoth petition of 85,000 signatures, and extensive media coverage. There were also some novel features, including a threat to strike if ICI continued to co-operate with NIREX and a proposal to pump water into the mine to render it unusable. By March 1984 the protesters made a major coup when ICI reversed their stance and declared that, 'Having carefully considered the implications that the proposal would have for its business, ICI has concluded that it would not be in the company's best interests and is therefore opposed to it'. This was the crucial factor in the subsequent abandonment of the Billingham option by the Government.

In Bedfordshire, opposition took on a different emphasis. Although a public campaign also called BAND (Bedfordshire Against Nuclear Dumping) was immediately formed and called public meetings, organised a petition and co-ordinated information, the lead was here taken by the County Council. A resolution supported by all political parties was passed within days of the announcement and expert consultants were employed to marshall the case against the proposal. A delegation was sent to the Minister and the County Council urged London Brick to apply for the Elstow site. When NIREX indicated its intention of taking soil samples from the site, the Council took out an interim injunction to prevent it. When that was refused both at the first hearing and on appeal, the Council applied a Stop Notice and Enforcement Action on NIREX. Although this could not operate until three days after sampling began, failure of the equipment on site rendered the sampling incomplete although NIREX claimed that it had obtained what it needed.

Within a year of the announcement, NIREX was confronted with powerful opposition in both areas. At Billingham it had failed to carry ICI and at Bedford it faced opposition from London Brick. Its failure to reveal alternative sites, to justify its site selection, and to match its proposals to government criteria, and its insensitivity to the weight of opposition encountered, might be thought to indicate a degree of incompetence, innocence or indifference. But, when the wider political aspects are considered, NIREX appears as the instrument of a policy designed to overcome a major obstacle for the industry. The Environment Secretary's action in absolving NIREX of any further political bad favour by withdrawing from Billingham may be judged a significant serious political backdown. And the decision to nominate further sites in addition to Elstow suggests that the Bedfordshire campaign had also achieved some success.

The wider political dimension

The debate over specific sites was conducted within a developing political conflict over the whole subject of land-based disposal. This conflict focuses on three issues. The first concerns the feasibility of the proposal. The argument that disposal of ILW 'has been validated as a safer option, by research and development, and by experimental or actual disposal and subsequent monitoring in several countries ' (DoE, 1984a, para. 28) was open to question. Research such as that undertaken by the NRPB was purely generic and was not intended to be applied to specific sites. There are neither experiments nor actual facilities for disposal of ILW as yet in the UK. Only two other countries, France and the United States (admittedly the two leading countries in nuclear power production), have experience of operating shallow land disposal facilities. The French facility at La Manche in Normandy has operated since 1969 although it is a different concept from that intended for Elstow. The American experience has been varied, as Chapter 8 shows. The site at Barnwell (South Carolina) is also a different concept and the two other operating sites at Beatty (Nevada) and Richland (Washington State) do not have concrete trenches. Three shallow repositories have already been closed as a result of leakage; the one at West Valley (New York State) actually overflowed. As a result the United

States has embarked on an extensive research programme before identifying further sites.

Three countries, the Federal Republic of Germany, Sweden and Finland, have specifically rejected shallow disposal, and, along with three others (Belgium, GDR and Switzerland), have opted for deep disposal solutions. Of these, Sweden is planning to construct storage chambers under the sea bed offshore, a very costly solution compared to that proposed for the UK (Blowers, 1986). Switzerland has specifically adopted a requirement that risk shall be reduced as far as is reasonably achievable within the state of science and technology, a far more stringent application of the ALARA principle than applied in the UK. There is, therefore, little evidence to substantiate the claim that the acceptability of land disposal has been widely demonstrated.

The view that there was a lack of comparable experience of shallow burial and inadequate research into disposal options was supported by the House of Commons Environment Committee, which reported its study of radioactive waste in March 1986 (HMSO, 1986a). The Committee was impressed by the evidence from Bedfordshire County Council which underlined that 'feeling runs high and that such feeling is based on reasonable scientific doubt' (Ibid., para. 87). If the lack of research into alternative options was to be justified on the basis that there was already sufficient foreign experience, 'then we must actually take heed of foreign experience' (para. 88). The Committee recommended that greater research into seabed options was needed and concluded that shallow burial was only acceptable for 'short-lived low-level wastes and must be fully engineered on a complete containment basis' (para. 99).

A second issue is the lack of information on what quantities and what types of waste are destined for the proposed land disposal facilities. 'The precise types of waste that can be accepted at each facility ... will depend on the quantities involved, the details of the eventual design of the facility, and the packaging of the waste' (DoE, 1984a, para. 28). In other words, the decision as to what to dispose of would be taken <u>after</u> the repositories were constructed. There has been considerable confusion over definitions of different types of waste. At the Sizewell Inquiry the DoE witness stated that, 'LLW, by our definition is that of which we can dispose ... '. (Sizewell 'B'

Inquiry Proceedings, Day 100, p. 12) inviting the conclusion that once a disposal route for ILW had been secured it could be redesignated as LLW. This prompted the comment during cross-examination that this method of categorisation, 'involves a slurring of wastes of one form of radioactivity from one category to another, and possibly in the public eye a consequent diminution of the seriousness of the problem of disposing of them' (Ibid., p. 17). The inadequacy of this definition led to a subsequent announcement by NIREX that it had reached an agreed definition of LLW based on radioactive content. The details of this are set out in a DoE document submitted to the Sizewell Inquiry in September 1984 (TCPA, 1984). The problem of definition and the notion of postponing decisions about the types of waste to be disposed of until the facilities are constructed leads to the understandable fears that Elstow and the other potential sites are designated for undisclosed types and uncertain quantities of waste.

On the question of classification of wastes, the House of Commons Environment Committee recommended that LLW should exclude any wastes with half-life of over thirty years, alpha-particle-emitting wastes, and toxic radionuclides (HMSO, 1986a). In its response to the Committee's Report the Government accepted the need for clear definition and agreed that shallow burial sites should receive only LLW in order to reassure the public. In consequence the short-lived ILW originally destined for shallow burial would, like the long-lived ILW intended for Billingham, be stored pending the development of a deep disposal facility.

The third issue surrounds the decision making procedures, which appear to have been unduly compressed. NIREX has claimed that there is a need for the new facilities, which it hopes will be operational by about 1990. Yet the DoE witness at Sizewell argued that, 'the heavens will not fall' if they are not available in the 1990s. 'This material is in store, it will add to the inconvenience and expense of looking after it, but it is safely in store now' (Sizewell 'B' Inquiry, Day 100, p. 23). In his original announcement the Secretary of State suggested that the investigatory works may require permission, but subsequently he made it clear that he would call in both the first stage of exploration and the subsequent application for development for his determination. Later, in January 1985, the Secretary of State decided that planning permission for the first stage of exploration would be granted by a Special Development Order laid before Parliament to be followed by a public inquiry 'which would examine all the alternative sites and the environmental assessment for each site' (Hansard, 24 January 1985). The SDO was laid before Parliament in May 1986 and passed by a majority of 78. Those voting against it included Conservative MPs with constituences including or close to some of the prospective dumps, and the Opposition, who argued the principle that the local communities had been denied the right to a local planning inquiry at the first stage.

Although there would be a major public inquiry when NIREX eventually applied for permission to develop a shallow repository, the precise nature of the inquiry was unclear. The Secretary of State for the Environment appeared to expect an inquiry into one of the sites. In his statement he said:

> I understand that the investigation of the four sites could take between 12 and 18 months. If any of the sites prove to be suitable, NIREX would at the same time be in a position to decide what proposals it wants to make the subject of a planning application, (Hansard, 25 February 1986).

Bedfordshire had intended to oppose the Government's shallow disposal strategy at the promised inquiries on exploratory investigation. Having been denied that opportunity the Council expected that, following the precedent set by Sizewell, the public inquiry on the planning application by NIREX would be wide-ranging and would include issues of principle as well as those which were site-specific. But there were signs that the terms of reference of the eventual inquiry would be limited. Already demands for a Planning Inquiry Commission to consider the wider aspects of the strategy for nuclear waste management had been rejected. And, following the Holliday Report, the DoE had conducted its own investigation into the Best Practicable Environmental Options (BPEOs) (HMSO, 1986b). This had concluded that the BPEO for 'most LLW and some short-lived ILW is near-surface disposal, as soon as practicable, in appropriately designed trenches' (p. 5). It was, therefore, clear that the land burial strategy was favoured by Government and that it would seek to legitimate it. If it succeeded then it could claim that the policy was established and therefore limit the debate at the inquiry to site-specific issues. It appeared that the political imperative for a solution to the radioactive waste problem led inexorably to a pragmatic and premature solution in favour of land disposal.

Alternative futures

For the present, sea disposal is not available although NIREX continues to pursue this as a possible future option for some wastes. It is also investigating methods of disposal under the sea bed though this is unlikely to divert much attention from the main effort which focuses on land disposal. Storage at existing sites has been rejected as both too costly and as a problem which should not be left for future generations to solve. Ironically, this is precisely the problem which will face future generations in the case of HLW and ILW.

The opposition to the Government's proposals for radioactive waste has scored some notable successes. The abandonment of Billingham, the announcement of alternative sites for shallow disposal and the exclusion of ILW from such sites represent significant concessions on the part of government. Paradoxically, removal of the ILW problem may leave one of the four named sites more vulnerable to development on the grounds that shallow disposal for LLW is already accepted practice at Drigg and elsewhere. Other, more fundamental, changes could occur. If the nuclear energy programme and reprocessing were to be phased out then the life of Drigg could be extended. Since decommissioned power stations will have to be left for at least 100 years to cool off it makes little sense to transport wastes elsewhere when they could be left at the power stations. Given the variety of possibilities, critics argue for 'a much broader, and more thorough and rigorous public investigation of all aspects of radioactive waste management strategy, including alternative disposal options, site selection criteria and reprocessing policy, before any public inquiries are held either into proposals for actual repositories or into test drilling in one or two selected sites' (TCPA, 1984, pp. 6-7).

Such a comprehensive approach is unlikely to prove acceptable either to the industry or the Government. The option they have followed confirms the proposition that it is political necessity rather than rational analysis that determines policy in this field. The selection of specific sites, the denial of an early inquiry, the assertion that nuclear expansion is justified, are all means whereby government and the industry have been able to control the direction of policy. But the decision making strategy raises some fundamental questions about accountability to this and future generations. It manifests an inequality of power, in which the interests of the nuclear industry enjoy prior access and close relationships with government, and which excludes local interests and restricts their ability to question the broad dimensions of policy. At best the strategy for waste disposal on inland sites is an example of incremental, pragmatic decision making. At worst it may be viewed as an attempt to legitimate a solution which favours powerful corporate interests in the face of public anxiety.

There is, however, now the possibility that the strategy may fail. Opposition forced the Government to retreat on sea disposal and on HLW disposal and to abandon Billingham. The four shallow burial sites appear to have been selected on grounds of expediency rather than what may broadly be termed rationality. The whole strategy for shallow burial appears questionable when evaluated against other options and the developments in other countries. Above all, growing disquiet about the record and policies of the nuclear industry, partly as a consequence of Chernobyl, may cause fundamental changes in policy which, at present, may not be foreseen.

Note 1

The dose equivalent is measured in millisieverts (mSv) and is a measurement of the amount of ionising radiation intercepted by various tissues, weighted by the risk of occurrence. A 'critical group' is that small group of people expected to be most exposed to the highest danger from radiation.

REFERENCES

ARNOTT, D. G. (1982) 'The end of the affair', PANDORA Newsletter, 10 January.

ASHBY, E. and ANDERSON, M. (1981) The Politics of Clean Air, Oxford: Clarendon Press.

BEDFORDSHIRE TIMES, 10 Nov. 1983.
BEDFORDSHIRE TIMES, 27 Nov. 1983.
BEDFORDSHIRE TIMES, 27 Feb. 1986.
BLOWERS, A. T. (1984) Something in the Air: Corporate Power and the Environment, London: Harper and Row.
BLOWERS, A. T. (1986) 'Sweden buries its nuclear waste problem', Geography 69, 260-64.
CEGB (Central Electricity Generating Board) (1982a) Feasibility Study of Shallow Land Burial of Specified Intermediate and Low Level Radioactive Wastes, study by Pollution Prevention (Consultants) Ltd., submitted to Sizewell B Inquiry, April.
CEGB (1982b) 'The disposal of low level and intermediate level solid wastes', Proof of evidence by R. FLOWERS (NIREX) at Sizewell B Inquiry, CEGB/P/21, November.
DoE (1983, 1984b and 1984c) Disposal Facilities on Land for Low and Intermediate-level Radioactive Wastes: Principles for the Protection of the Human Environment, Draft October 1983, interim version submitted to Sizewell B Inquiry, September 1984, and final version issued December 1984.
DoE (1984a) Radioactive Waste Management: the National Strategy, July.
DUNSTER, J. H. (1958) 'The disposal of radioactive liquid wastes in coastal waters' Paper P/297 to UN Conference on the Peaceful Uses of Atomic Energy, Geneva. Proceedings, 18, 390-399.
FALK, J. (1982) Global Fission, Oxford: Oxford University Press.
FoE (Friends of the Earth) (1984) The Gravedigger's Dilemma, prepared by Rene Chudleigh and William Cannell for FoE, London: FoE Publications.
GOWING, M. (1964) Britain and Atomic Energy 1939-1945, London: MacMillan.
GOWING, M. (1974a) Independence and Deterrence: Britain and Atomic Energy 1945-1952, Volume 1, Policy Making, London: MacMillan.
GOWING, M. (1974b) Volume 2, Policy Execution, London: MacMillan.
GREEN, H. P. (1982) 'The peculiar politics of nuclear power', Bulletin of the Atomic Scientists, 38, 10, 58-64, December.
HANSARD, 25 October 1983.
HANSARD, 3 May 1984.
HANSARD, 24 January 1985.
HANSARD, 25 February 1986.

HAWKINS, K. (1984) Environmental Policy and Enforcement, Oxford: Clarendon Press.
HMSO (1959) The Control of Radioactive Wastes, Cmnd. 884, London: HMSO.
HMSO (1976) Nuclear Power and the Environment, Sixth Report of the Royal Commission on Environmental Pollution (Flowers Report), Cmnd. 6618, London: HMSO (September).
HMSO (1979) The Control of Radioactive Wastes: a Review of Cmnd. 884, HMSO: London (September).
HMSO (1982) Radioactive Waste Management, Cmnd. 8607 (White Paper), London: HMSO (July).
HMSO (1984) Report of the Independent Review of Disposal of Radioactive Waste in the Northeast Atlantic (Holliday Report), London: HMSO (December).
HMSO (1986a) Radioactive Waste, first report of the Environment Committee 1985-6, House of Commons, March, London: HMSO.
HMSO (1986b) The Assessment of Best Practicable Environmental Options (BPEOs) for the Management of Low and Intermediate Solid Radioactive Wastes, Radioactive Waste (Professional) Division of the Department of the Environment, March, London: HMSO.
JAY, K. E. B. (1952) Harwell, London: HMSO.
JAY, K. E. B. (1955) Atomic Energy Research at Harwell, London: Butterworth.
JAY, K. E. B. (1961) Nuclear Energy, Today and Tomorrow, London: Methuen.
MEDVEDEV, Z. (1976) 'Two decades of dissidence', New Scientist, 72, 264.
MEDVEDEV, Z. (1977) 'Facts behind the Soviet nuclear disaster', New Scientist, 30 June, 761-764.
MEDVEDEV, Z. (1979) Nuclear Disaster in the Urals, London: Angus and Robertson.
NIREX (Nuclear Industry Radioactive Waste Executive) (1983) The Disposal of Low and Intermediate-level Radioactive Wastes: the Elstow Storage Dept. A Preliminary Project Statement, Harwell: NIREX (October).
NRPB (National Radiological Protection Board) (1982) Radiological Protection Aspects of Shallow Land Burial of PWR Operating Wastes, prepared by HILL, M. D., and PINNER, A. V., Didcot: NRPB (December).
NRPB (1984) An Assessment of the Radiological Protection Aspects of Shallow Land Burial of Radioactive Wastes, prepared by HEMMING, C.

R., HILL, M D. and PINNER, A. V., Didcot: NRPB (April).
PRINGLE, P. and SPIGELMAN, J. (1983) The Nuclear Barons, London: Michael Joseph.
RHODES, C. (1981) Inspectorates in British Government, London: Allen and Unwin.
SCRAM (Scottish Campaign to Resist the Atomic Menace) (1980) Poison in Our Hills: the First Inquiry on Atomic Waste Burial, Edinburgh: SCRAM.
Sizewell B Inquiry (1984a) The Human Consequences of Exposure to Ionising Radiation, Proof of Evidence, SSBA/P/8 prepared by Dr Rosalie Betell, August.
TCPA (Town and Country Planning Association) (1984) Disposal Facilities on Land for Low and Intermediate Level Radioactive Wastes. Draft Principles for the Protection of the Environment. Comment by the Town and Country Planning Association, London: TCPA (April).
TRABALKA, J. R., EYMAN, L. D. and AUERBACH, S. I. (1979) Analysis of the 1957-58 Soviet Nuclear Accident, Publication No. 1445, Environmental Sciences Division, Oak Ridge National Laboratory, Oak Ridge.
TUSIU (Trade Union Studies Information Unit) (1980) The Nuclear Triangle: the Miners Report on Waste Dumping in the Cheviots, Newcastle: (October).
WEIL, G. L. (1955) 'Hazards of nuclear power plants', Science, 121, 3140, 4 March, 315-317.
WILSON, C. L. (1979) 'Nuclear energy: what went wrong?' Bulletin of the Atomic Scientists, Chicago, 35, 6, 13-17, June.

Chapter Seven

INSTITUTIONAL ASPECTS OF SITING NUCLEAR WASTE DISPOSAL FACILITIES IN THE UNITED STATES

John Cameron Stewart and W. Clark Prichard*

* The views expressed here are solely those of the authors.

Introduction

This chapter focuses on what may be called the 'institutional' issues facing nuclear waste disposal policy makers. These are distinct from technological considerations of engineering and geology, and include socio-economic, financial, land use and, especially, political factors. While the technological (geological and engineering) aspects of siting nuclear waste facilities are critical, the institutional issues (especially political) are also important. Institutional issues must be resolved before a nuclear waste facility becomes operational, regardless of how technically suitable a site may be for nuclear waste disposal. The US Nuclear Regulatory Commission's (USNRC) regulations for siting nuclear waste facilities include both technical and institutional requirements - listed in Federal Codes 10 CFR Part 60 for high level waste (HLW) and 10 CFR Part 61 for low level waste (LLW): the US does not have an intermediate level in its waste classification. Although their emphasis is on meeting technological requirements in siting nuclear facilities, the USNRC regulations recognise the significance of land use, land ownership, financial assurance, demography, and other institutional issues related to site selection, operation and closure (10 CFR Parts 60.121-60. 122 for HLW and 61.14-61.15 and 61.50 for LLW).

Because institutional considerations (especially political) are important in site selection, local community acceptance of nuclear waste disposal facilities is essential to the site selection process. In the US as in the UK, it generally will be extremely difficult to overcome local opposition to siting a new nuclear waste facility. However, largely due to a combination of favourable geographical and economic factors at certain locations in the US, there are several communities which have expressed a positive interest in having a nuclear waste facility. These favourable factors are found in small communities in areas of very low population density, where uranium mining and/or milling operations have recently closed. These community acceptance issues will be discussed in more detail later in this chapter.

Overview of Nuclear Waste Issues in the U.S.

Commercial (non-defence) HLW currently includes spent nuclear fuel and certain fuel reprocessing wastes. The Nuclear Waste Policy Act of 1982 gives the NRC authority to add highly radioactive wastes requiring permanent isolation to this category, and the NRC is currently considering ways to implement this authority. LLW is presently defined as anything not classified as HLW, but current legislative initiatives are likely to narrow this definition considerably.

There is an urgent need to find disposal sites for both HLW and LLW. However, both LLW and HLW site and management systems involve many institutional issues, which must be resolved before site selection can occur. These include political, social-economic and land use impacts in the vicinity of proposed waste sites, and along transport routes leading to these facilities. In the US, land use controls for private lands are a state or local government responsibility, and thus such controls are highly variable across the nation. Some communities have strong land use controls and planning programmes, while others (especially in sparsely populated areas) take a generally *laissez-faire* attitude towards land use control on private property.

The US Department of Energy (USDOE) has the responsibility of selecting and operating a site for the first HLW repository (and second if necessary). The USNRC has the authority to license HLW repositories contingent upon their meeting health and safety requirements. Three potential HLW sites have been selected for investigation: Deaf Smith County, Texas (Salt geologic formation); Richland, Washington State (Basalt geologic formation); Yucca Mountain, Nevada (Tuff geologic formation), Figure 7.1.

Figure 7.1 LOW LEVEL WASTE DISPOSAL SITES IN THE US AND PROPOSED HIGH LEVEL WASTE DISPOSAL SITES

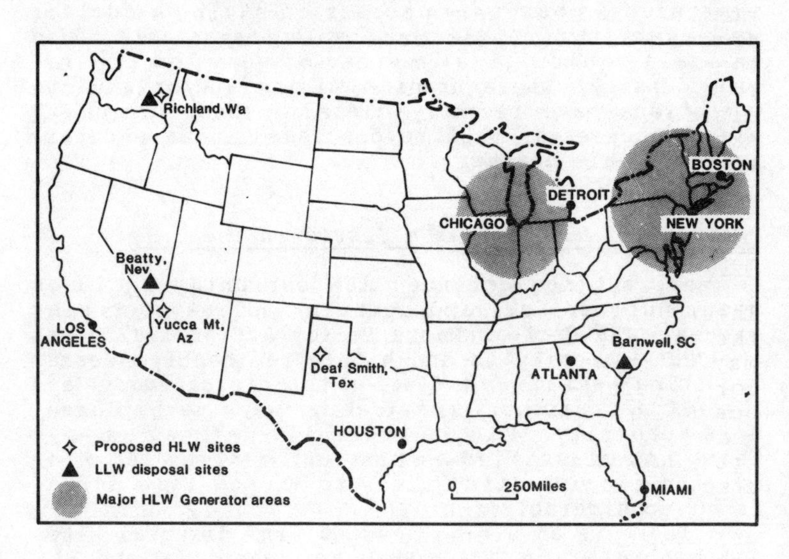

The need to find new disposal sites is especially critical for LLW, where a state or group of states may form geographical units that exclude LLW shipments from sections of the country. Three existing LLW sites are at Barnwell, South Carolina; Beatty, Nevada; and Richland, Washington, Figure 7.1. LLW management in the US is being dramatically changed by passage of the Low-Level Radioactive Waste Policy Act of 1980. The Act places responsibility for disposal of these wastes upon the states, with the Federal government assuming a lesser role. It encourages states to form regional groups ('compacts' in the language of the Act) to establish and manage LLW disposal sites to dispose of LLW generated within the compact.

The 1980 Act allows compacts to exclude the shipment of LLW from states outside their compacts as an incentive to form such compacts and to establish new LLW sites.

Nine compacts, representing most sections of the country, have been or are being formed. Some states, such as Texas, have elected to go it alone and will handle all of their own LLW by a disposal site within the state. One problem, however, with most of these compacts is that the site selection process is still far from being settled. Most compacts have no existing or approved new LLW disposal sites. Even in compacts with existing LLW disposal sites, there is the question as to which state will handle LLW disposal after the existing sites reach capacity.

Organisation and financing of nuclear waste management

The US HLW management programme is presently under the control of the USDOE. However, the programme may be transferred to the control of a new public corporation created for the express purpose of operating it. The public corporation would be a quasi-governmental organisation, under Federal control but outside the traditional Federal bureaucracy.

The HLW programme is to be funded from a fee or tax on nuclear-generated electricity. The present fee is one-tenth of one cent for each kilowatt-hour generated. However, it can be altered as needs arise. This financing arrangement has widespread support. Since nearly all of the civilian HLW which is generated comes from nuclear reactors, the arrangement fits the social equity criterion of 'benefit/burden concordance', e.g. those who directly benefit from waste-generating electricity production - the electricity users - will bear the cost of safely disposing of the waste. (For a complete discussion of equity issues, see Kasperson et al., 1984.)

The fund from the kilowatt-hour fee will support all of the expenses associated with storing, transporting, packaging, and placing HLW in a safe place - most probably in HLW repositories in areas having suitable geological characteristics.

The fund is also supposed to finance decommissioning of repositories and to provide for the long-term maintenance and surveillance which is planned as part of a period of institutional control over the repository. (Requirements do not assume any control beyond 100 years although the waste is hazardous for a much longer period.) This period includes:

1) control and management of a closed HLW facility (security, routine maintenance, and maintaining a technical staff at the site):
2) monitoring (observing seismic, thermal, and radiological conditions to detect changes):
3) information transfer (recording data about the site) (USDOE, 1980).

Financing for the post-closure phase of the HLW facility is more uncertain than for the operational phase. The costs of post-closure activities are difficult to forecast since they will be occurring many decades into the future. Also, it is not clear whether there will be enough revenue from the kilowatt-hour fee, since technological changes could make nuclear-generated electricity obsolete during the hundred years following closure. Revenues from the kilowatt-hour fee could dwindle to nothing while at the same time costs of monitoring the sites may be rising.

Even at the preliminary stage of the formation of regional LLW compacts, problems have been encountered. There have been disagreements over procedures for selecting a disposal facility site and for managing a regional waste disposal system, and there have been jurisdictional problems. It appears that nine or more compacts will eventually emerge. Each will eventually serve a specific geographical area. Whereas all operating LLW disposal facilities have to date used shallow land burial techniques, the compacts and individual states are considering a wide range of other options, including abandoned mines and engineered disposal facilities.

The next stage for most compacts will be to decide on disposal technology and to select actual sites. Once a facility is developed and goes into operation, managing it will be the major concern of each LLW compact.

The waste management system of each regional compact will be funded by a fee based on quantity of LLW actually disposed of. Waste generators will bear the costs of disposal. Most of the existing disposal sites charge a fee which is based on volume and class (radioactivity level) of waste. This fosters volume reduction practices by waste generators. It is possible to reduce LLW into much smaller volumes, thus economising on transport and disposal costs.

The differences between the organisation of

LLW and HLW disposal systems result from several factors:

1. Relative hazard of the waste: HLW is extremely radioactive, requiring extensive precautions to safeguard public health and safety - including expensively engineered facilities for long term isolation. LLW is generally far less radioactive and allows for near-surface disposal provided the siting requirements of 10 CFR Part 61 are met.

2. Institutional and historical factors: HLW has always been a Federal responsibility in the US. The Federal government has a long history of managing defence-related high level radioactive wastes. LLW management has in the past been a responsibility of both Federal and State authorities, with a tendency in the very recent past toward more State responsibility.

3. Relative quantities of wastes generated: There is far more LLW. The annual rate of generation averages about 3 million cubic feet. Because of this, transport costs are a much more important consideration in managing LLW than in HLW. The formation of regional LLW compacts should greatly reduce transport costs for LLW disposal.

Because of these factors, Federal control over HLW management was established. This required extreme precautions and expensive technology. The relatively low volume of HLW and the stringent requirements for a safe disposal site resulted in the Federal government concentrating its efforts on building just one, or possibly two, geological repositories.

LLW could be disposed of much more readily and less expensively, and states did not feel as limited by resources or experience in dealing with disposal technology for LLW management as compared to HLW. The lower hazard level and higher volume of LLW resulted in economic factors playing a more important role. Instead of just one or two disposal sites, the US could have as many as 12 LLW sites. This is because transport costs are a major factor, and near-surface disposal sites adequate for LLW disposal are more readily available than sites for HLW (requiring deep underground disposal), and less expensive to develop.

Land use issues

Siting new nuclear disposal facilities is a strongly political matter, not just a technical one (USDOE, 1985, p. 6).

Local land use controls may play a major role in the matter, especially for LLW, in parts of the US. Citizens in communities which have relatively strong land use controls will probably attempt to use them to impede a proposed disposal site in their community, regardless of any known public health or safety considerations. Even communities without strong land use controls may attempt to block nuclear disposal sites by enacting specific ordinances. Although several LLW state compacts have been formed, no compact with an acceptable site was approved by Congress by the original January 1986 deadline. This deadline has been tentatively extended to 1993.

Community acceptance

While most communities are likely to oppose the siting of a nuclear waste facility (HLW or LLW) in their area, some, such as Edgemont, South Dakota, and Naturita, Colorado, may accept such facilities. Potential acceptance is related to geographical and economic factors. The arid US West has large areas of very low population density. Such areas already contain most of the uranium ore deposits and mining and milling operations but most are now closed. Hence new employment opportunities provided by opening nuclear waste disposal facilities may be attractive to communities familiar with the nuclear fuel cycle. These communities, near existing or closed uranium mining or milling operations, may not be concerned about the potential of additional environmental or radiological hazards from nuclear waste facilities, provided adequate measures are taken to protect the health and safety of the general population (as required by USNRC regulations).

Community acceptance is also directly related to economic benefit to be gained. The potential of 1,000 - 2,000 jobs for a high level repository could have a very positive impact on a sparsely populated community, especially where there is high unemployment. However, it is important that the community should believe that it will directly benefit from the repository, through the employment of local workers and through increased local spending by repository personnel.

According to a recent socio-economic impact study, there is less opposition to nuclear plants from communities that have the following combination of characteristics:

> reliance on established government procedures; political values which belittle protest; a prevailing, pro-growth attitude; an important community leadership role played by the business community; and perceived economic benefits from plant construction and operations (Chalmers *et al*., 1982).

Communities whih have similar values will probably also be receptive towards a nuclear waste facility, especially where the nuclear industry is currently or was previously in existence.

The small communities of Naturita, and Edgemont generally support having a LLW disposal facility in their areas. They have many of the characteristics described above, and have also had unemployment problems due to the closing down of nuclear facilities. '... Naturita, Colorado, a community located in a uranium mining region of the US, actively solicited consideration as a potential site for a low-level radioactive waste disposal facility'. In addition, 'Edgemont, South Dakota, has also pursued the possiblity of hosting a low-level radioactive waste disposal facility' (USDOE, 1985, 65).

Many of these community acceptance factors apply also to the Richland, Washington area, where a HLW repository may be sited. In the past few years this area has experienced a marked increase in unemployment due, in a large part, to cancellation of the construction of several large nuclear reactors. As the nuclear industry has been very significant to the Richland economy since World War II, it is likely that the local community will generally support a HLW repository, especially since the economy is currently depressed. However, local farmers are concerned about possible contamination of the Columbia River, which is the chief source of irrigation water in this semi-arid eastern section of Washington. Environmental and geological concerns at both the Naturita (LLW) and Richland area (HLW) sites must be assessed further.

By contrast, the Deaf Smith County, Texas, community sees no overall economic advantages in having a HLW repository in its area. On 11 February 1985, a Congressional hearing was held at Hereford, Texas, near the potential HLW repository site. It was apparent that there was no local public support for the HLW repository. The principal concern expressed by local and state governmental officials and the general public was the possible contamination of the underground Ogallala Aquifer. This aquifer is very large (176, 940 square miles) and is of extreme economic importance to the agricultural community in this semi-arid region. Deaf Smith County is in a major seed and crop producing region, irrigated largely by the Ogallala Aquifer. People are concerned that the locally-grown agricultural products would be difficult to market elsewhere due to public concern about contamination (real or imagined). According to Jim Hightower, Texas Commissioner of Agriculture, ' ... we know that even without a leak in this (HLW) facility, for example, the public perception of food grown over nuclear waste is going to have an immediate and long term negative effect on (local) agricultural sales' (US Congress, 1985).

Transport issues

Another major siting factor that must be addressed is the problem of transport and its impact on land use adjacent to nuclear waste shipping routes. It is anticipated that a HLW repository will be located in the Western US, while most HLW is generated in the Eastern US. This will give rise to long distance HLW transport, generally by truck along interstate highways. This means that the HLW vehicles will come close to large metropolitan poulation centres along the shipment routes.

It is important to stress the fact that there has been extensive research by USNRC into HLW containers. They are designed to survive high speed impacts without release of radioactivity to the environment. However, the radioactive waste transport has already created debate in many communities along projected routes. Non-technical issues will probably have a major impact in selecting transport routes. According to a study prepared for the NRC:

> transportation issues are a relatively visible component of the nuclear energy controversy. Although the number of transportation incidents which have occurred has been small, and the consequences slight, the political and legal attention given to transportation is likely to increase. These impacts may ultimately prove more significant in decisions regarding the transportation of radioactive materials than will strictly technical concerns. (Findlay et al., 1980).

Even though the risk of exposure of radioactive waste to the population is extremely small, routes should avoid large population centres, as an extra safety measure and also to avoid major delays in shipments due to accidents or road and bridge construction. In addition, areas of extreme weather should be avoided. The Governor of Colorado expressed concern about the possible nuclear waste transport from East to West through Denver and the Rocky Mountains. He feared that the routes would often be closed during the winter due to heavy snow, necessitating a more southern routing of HLW.

Hence, the geographical factors which should be considered in evaluating alternative routes for radioactive waste shipment are: proximity to population and large urban centres; quality of highways (interstate highways should be used as much as possible); terrain; climate.

Local economic impacts

The three existing LLW facilities in the US use shallow land burial techniques. Two of these, Richland, Washington, and Beatty, Nevada, are in sparsely populated areas in the arid Western US. The one humid site, located near Barnwell, South Carolina (1980 population for Barnwell County, 19, 868) is also the largest facility, occupying 250 acres and employing approximately 260 people (see also Chapter 8 of this volume). All LLW sites are government-owned land, either Federal or State. The actual site facilities are contractor operated - the contractors being regulated by State and Federal licences.

Because of the relatively small size of the LLW facilities, especially at Richland and Beatty, there are not the socio-economic impacts on the local area which normally occur with a nuclear power plant or other large construction project. The impact of the Barnwell facility on its local community is somewhat more significant than for the other two smaller LLW facilities. However, the socio-economic impact of the Barnwell LLW site is still very small in comparison to what a full-scale nuclear plant would have had.

There has been no major impact on employment and few significant disruptions of the local economy and social structure around all three LLW facilities. There are no special impacts, such as an immediate demand for new housing and schools, which are often associated with constructing and operating a new nuclear facility. Also, LLW waste is not as hazardous as HLW, so the health and safety threat, real or perceived, of a catastrophic accident is not a major concern.

The mostly rural nature of present US sites suggests that effects on the marketability of local agricultural products could be an important issue. The sites at Richland (Washington) and Barnwell (South Carolina) are in areas where agriculture is important, but there does not appear to be any evidence that local agriculture has been affected.

Impacts on property values, residential or commercial, are another potentially important concern. Here again, these are not readily apparent in the present LLW sites.

It remains to be seen whether the next generation of sites, being built to accommodate regional compacts, will have similarly minor socio-economic impacts on surrounding areas. Newly developed facilities may be larger, may employ more people at either the construction or operational phase, and may use quite a different technology for waste handling and disposal. In addition, some may be in less remote areas, where larger populations would be concerned about the facility.

Unlike the LLW facilities, the first geological repository being planned for the United States HLW programme is an immense undertaking. The facility would occupy several thousand acres and employment would range from one to three thousand during peak construction, and from one to two thousand during operation (USDOE, 1979). The magnitude of the project, coupled with the fact that likely sites are in very rural (and in some cases, remote) areas, practically ensures that the economic impact on the surrounding area will be significant. There will almost certainly be a large influx of population, with a concomitant increase in the demand for local community services and housing. This situation has resulted in major problems in some instances, particularly in remote areas of the Western US associated with mineral development. However, recently there have been some examples of the ability of local areas to absorb this type of economic impact without undue strain.

A HLW repository may have an important economic effect on a community for whereas a power plant requires several thousand construction workers, it only needs several hundred for operation. This has a tendency to produce a boom-and-bust cycle. By contrast, the HLW waste repository would require far more operating personnel relative to the construction labour force, which could result in a more stable impact.

The geological repository will be an almost permanent feature of the local community since it is planned to isolate wastes for a thousand years or more. This is quite unlike other major development projects, which have an expected lifetime measured in years or decades.

Among the local economic effects which might be expected to occur due to the special nuclear or potentially hazardous aspects of a HLW facility are impacts on the marketability of local agricultural products, on property values, and on tourism or recreational industries. These would be more or less important depending upon the specific site under consideration. For example, one candidate site, Bryce Canyon in Utah, is near a major National Park. There has been concern about the possible effect of a HLW repository on tourism in Utah. Another, Deaf Smith County in Texas, is a major agricultural area, and there is a widespread belief among local residents here that the agricultural economy will be severely disrupted by the location of a HLW site. Indeed, severe declines in the price of agricultural land, which are attributed to this concern, have already occurred. It is important to note that some impacts may be offsetting; demand for residential property could be affected positively by an influx of workers and negatively by any unwillingness to reside near the facility.

The site which is selected for the first US HLW facility is virtually assured of some advance financial compensation for any anticipated socioeconomic impacts by the terms of the Nuclear Waste Policy Act of 1982. Among the impact mitigation terms of the legislation are provisions for funding states to conduct their own studies of likely impacts - funding designed to mitigate the impact of the development of a repository - and an annual grant to compensate State and local governments for the tax revenues lost due to the repository land and activities being non-taxable. All of this funding would come from the kilowatt-hour fee.

Conclusion

This chapter has dealt with the institutional issues associated with disposal of nuclear waste in the US. We believe that these institutional problems must be resolved, no matter how technologically well suited a site may be for disposal, before site selection may take place.

We have also pointed out that the geography of the US, with its large arid regions of very low population density, contributes to the institutional acceptability of nuclear waste disposal. Economic factors, especially in sparsely populated areas where the uranium mining and milling industry has ceased operation, also weigh on the acceptability of nuclear waste to local communities. This acceptability will be highest where there are existing nuclear facilities and/or facilities which are closed - thus creating unemployment - especially where alternative economic opportunities are few.

REFERENCES

CHALMERS, J., PIJAWKA, D., BRANC, K., BERGMANN, P., FLYNN, J. and FLYNN, C. (1982) Socio-economic Impacts of Nuclear Generating Stations: Summary Report on the NRC Post-Licensing Studies, Washington DC: US Nuclear Regulatory Commission, NUREG/CR 2750, 116.

FINLAY, N. C., ALDRICH, D. C. et al. (1980) Transportation of Radionuclides in Urban Environs: Draft Environmental Assessment, Washington DC: US Nuclear Regulatory Commission, NUREG/CR 0743, 12.

KASPERSON, R. E., DERR, P. G., KATES, R. W. (1984) 'Confronting equity in radioactive waste management: modest proposals for a socially just and acceptable program', in PASQUALETTI, M. J. and PIJAWKA, D. K. (eds.) Nuclear Power: Assessing and Managing Hazardous Technology, Boulder, Colo: Westview Press.

US CONGRESS (1985) The Socio-economic Effects of a Nuclear Waste Storage Site on Rural Areas and Small Communities, Rural Development Subcommittee of the Committee on Agriculture, Nutrition and Forestry of the US Senate, Washington DC.

USDOE (US Department of Energy) (1979) Draft Environmental Impact Statement: Management of Commercially Generated Radioactive Waste, Washington DC: USDOE, 1, 3.1.133.

USDOE (US Department of Energy) (1980) Final Environmental Impact Statement: Management of Commercially Generated Radioactive Waste, Washington DC: USDOE, 1, 3.49-3.50.

USDOE (Department of Energy) (1985) Use of Compensation and Incentives in Siting Low-Level Radioactive Waste Facilities, Washington DC: USDOE, 6.

Chapter Eight

BURIED FOR EVER? THE US EXPERIENCE OF RADIOACTIVE WASTE DISPOSAL

Marvin Resnikoff

CURRENT POLICY

The United States is the largest producer of radioactive wastes, and has considerable experience with shallow disposal methods. This experience has been mixed, with major problems of leakage and contamination occurring in those sites in the wetter eastern part of the country. As a result three sites have been closed down, and there is concern about the potential hazards at the remaining eastern site at Barnwell, South Carolina. This chapter, written from the perspective of the national environmental pressure group, the Sierra Club, analyses the geological and technical problems at each of the sites, and suggests the lessons that can be learned from this experience.

As the previous Chapter (7) has pointed out, the regulations for constructing and licensing nuclear waste disposal sites are promulgated by the US Nuclear Regulatory Commission (NRC). These must accommodate the environmental standards of the US Environmental Protection Agency (EPA), which have been agreed for high level wastes and are being developed for low level wastes.

High level waste is essentially irradiated nuclear fuel, the liquid or solidified fission products from processing nuclear fuel to extract plutonium, and whatever else the NRC classifies as high level. Under the Nuclear Waste Policy Act of 1982, the Federal government is responsible for the safe disposal of high level waste. The Department of Energy (DOE) is investigating potential sites in the south and west for siting a deep underground repository, which it is hoped will be operating by the end of the century. This investigation poses the question of whether the earth, 1000 to 3000 feet underground, can contain radioactivity for one million years or so without releasing it, and highlights the problem of transporting high level waste over large distances, affecting many communities *en route*. From the viewpoint of environmental bodies like the Sierra Club there is strong support for finding a permanent home for these highly radioactive materials, but they consider that the schedule proposed by Congress is much too ambitious to carry out a scientifically responsible programme. In any case, procedural obstacles will delay the operation of the repository until well after the year 2000. Meanwhile, as in other countries, these wastes remain in storage pending final disposal.

Under the Low Level Radioactive Waste Act (LLRWA) of 1980, Congress passed the responsibility for managing low level wastes to the States. Congress encouraged the States to form regional agreements, or 'compacts', to find common regional waste facilities. The LLRWA set a deadline of January 1986 for each region to locate and operate a waste facility. As Stewart and Prichard observed in Chapter 7, no region or state has met this deadline. There is resistance in the northeast and midwest because of the poor past performance of radioactive and toxic chemical landfills in these regions. Although there are few problems at the western sites, they are far from major waste generators, and this greatly increases transport costs and associated potential environmental impacts.

Policy for low level (which in the US incorporates much of the UK's 'intermediate level' category) waste disposal has evolved further in the US than in Britain, where, as Chapter 6 has shown, the Government's attempt to legitimate a shallow burial strategy has encountered considerable resistance. Given the continental scale of the US, land burial is geographically the most suitable option. The methods of disposal adopted in the early landfill sites in the US were primitive, and have more in common with the British experience at Drigg near Sellafield than with the engineered repository proposed by NIREX for Elstow and other sites in the UK. The NRC's regulations are intended to overcome the problems of management of future sites. Sites must have sufficient depth to the water table, be well drained and not subject to erosion. Disposal regulations require strong containers and emplacement which minimises voids. Buffer zones are required around the site. But, the regulations are also necessarily vague. For example, phrases such as 'to the extent practicable' are applied to the elimination of water, while the time period for preventing infiltration and for post closure control is set at 100 years although some wastes will remain hazardous for much longer. As in the UK, decisions will require interpretation of these regulations on a site-specific basis, which places reliance on experts and may lead to different standards being adopted for each repository. Consequently, environmental groups urge the identification of alternative methods of radioactive waste management, including long term storage and the reclassification of some longer-lived hazardous 'low level' wastes to high level for eventual disposal in the Federal repository when it becomes available.

WHAT IS 'LOW LEVEL' WASTE?

Low level waste, the subject of this chapter, is everything that is not high level, so it is defined by what it is not. It includes a broad spectrum of materials, ranging from mildly contaminated clothing and test tubes to intensely radioactive materials (see Chapter 7 for definition). Some 'low level' waste has greater specific radioactivity than liquid high level waste at Federal reprocessing plants.

Low level waste comes from nuclear reactors, medical and research institutions and industrial and government institutions. Reactor waste accounts for 24 per cent of the radioactivity sent to radioactive landfills (DOE, 1979). Over 95 per cent of the radioactivity in reactor waste is in irradiated components and reactor coolant purification media, that is, sludges and resins (Sierra Club, 1984). The dominant radionuclides are cobalt-60 and caesium-137, with half-lives of about five and 30 years respectively. An average reactor produces about 500 curies (Ci) of caesium-137 per year: eighty operating reactors therefore produce about 40,000 Ci per year, which are sent to radioactive landfills. The more voluminous dry reactor wastes are slightly contaminated, comprising only five per cent of the radioactivity, and can be greatly compacted.

Industrial waste accounts for 73 per cent of the low-level radioactivity going to landfills. Into this category falls the material from two large producers of isotopes for medical and research purposes: New England Nuclear (Massachusetts) and Union Carbide (New York), which respectively account for 24 per cent and 15 per cent of the total low-level waste radioactivity. Industrial waste also arises from radionuclides used by well drillers and radiographers.

Institutional waste, from hospitals and research institutions, accounts for about one third of the volume of waste going to radioactive landfills, but less than one per cent of the radioactivity. Medical waste is dominated by the short-lived technetium-99 , while research waste consists of carbon-14 and tritium, with half-lives of 5000 years and 12 years respectively.

COMMERCIAL RADIOACTIVE LANDFILL EXPERIENCE

Figure 8.1 LOCATION OF THE SIX COMMERCIAL RADIOACTIVE LANDFILLS IN THE US

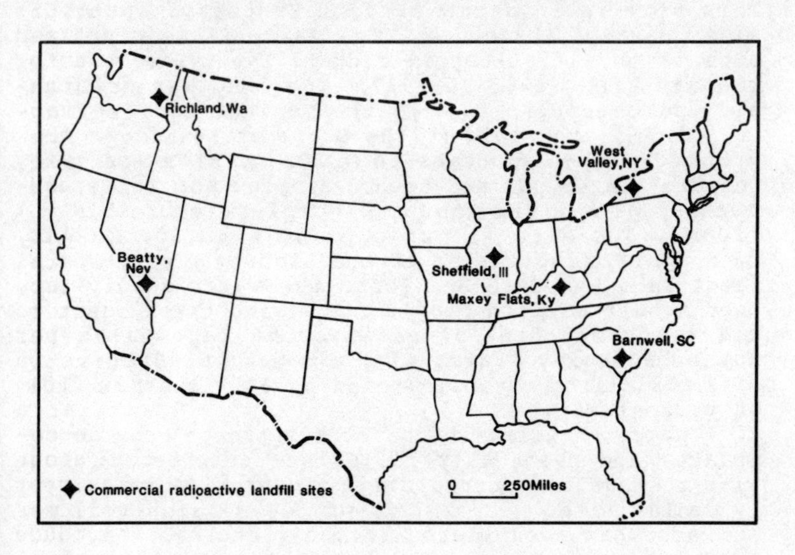

Of six commercial radioactive landfills (see Figure 8.1) which have operated in the US, three are now closed because of problems. These are at Maxey Flats, Kentucky; West Valley, New York; and Sheffield, Illinois. All three have suffered water infiltration into trenches, subsidence of trench covers, and erosion. At each site, radioactivity has migrated and expensive remedial actions are continuing. The major operating radioactive landfill, at Barnwell, South Carolina, is in a high rainfall area. It has not had build-up of radioactive leachate because of the porous, sandy trench bottom, which allows radioactive water to drain out of it. The other operating sites, in Beatty, Nevada and Richland, Washington, both in semi-arid regions, have apparently not had the same problems as at the more humid sites. The operational experiences at each of these sites illustrates the range of problems associated with shallow buried facilities for disposing of low level radioactive waste.

Maxey Flats

This site is in north-eastern Kentucky, about 65 miles east of Lexington. The Maxey Flats disposal site is on a flat-topped ridge. The 280-acre site operated from 1963 to 1977. The land was originally purchased in 1962 by the Nuclear Engineering Co. (NECO), then sold to the State of Kentucky, who granted a 25-year lease to NECO. Wastes are disposed of in 48 unlined trenches, plus hot wells and special pits. The annual precipitation is 110 to 120cm. The site is cut by, and sits 300 to 400 feet above, tributaries of the Licking River, which feeds into the Ohio and forms the water supply for over a half million people. Four and three quarter million cubic feet of radioactive material were buried at Maxey Flats, with a total radioactivity of 2.25 million curies, including 64 kilograms (kg) of plutonium.

Water entering the permeable trench caps accumulates and eventually fills the trenches, mixing with the burial materials and becoming radioactively contaminated. By 1972 over one million gallons of water had accumulated in the trenches. In 1973 an evaporation system was installed. Trench water was pumped into tanks: it later evaporated, and the resulting sludges were stored in a tank onsite. While the site operators claimed that subsurface migration of plutonium was impossible, the EPA found plutonium approximately three feet deep in core drilling samples. Plutonium was also found in surface soil, monitoring wells and drainage streams. In 1975 the EPA, following up on a State report on the migration of plutonium, concluded:

> The burial site was expected to retain the buried plutonium for its hazardous lifetime, but the plutonium has migrated from the site in less than ten years. If 100 per cent retention of a waste ... is the goal of shallow land disposal, continued burial of plutonium in humid climates using present waste forms, containers and trench construction methods, will not achieve the goal (Meyer, 1975).

Subsurface migration of radionuclides other than tritium has not been extensive compared to trench water concentrations. Tritium, probably from the evaporator plume, was detected in domestic well water. Milk samples, from cows within 3.1 km of the site, and drinking water from Rock Lick Creek, showed higher tritium levels. In 1977 the detection of cobalt-60 and manganese-54 in a seep in the newly opened trench 46 led to closure of the Maxey Flats landfill. The State of Kentucky purchased the lease rights from NECO for $1.2m, and now maintains the site. As the trench caps subsided, about two million gallons of water per year began entering the trenches. The evaporator processed 1.2m gallons, its maximum capacity, and an excess of 800,000 gallons remained in the trenches and on-site tanks. Finally, as a temporary remedial measure, in November 1981 the State covered three quarters of the trenches with PVC. The polymer membrane was covered with a geotextile material for structural strength and 45cm of earth to protect against weathering.

The principal pathway for water inmigration appears to be through the trench caps. Since the site is on a plateau, subsurface migration is probably minimal. With the plastic cover installed, the annual one metre rise in trench water was reduced to 0.3 metres, but not eliminated. Apparently small rocks and the movement of heavy equipment over the membrane created holes and pathways for water inmigration. Additional settling of the trench cover due to the deterioration of waste containers also degraded the membrane (Mills, 1983). Even twenty years was not sufficient time, apparently, for the cover to settle completely. Though water migration through the trench caps is the primary mode of trench recharge, subsurface migration still exists, through the more permeable sandstone strata. Pumping in one trench causes drawdown in the other trenches, indicating that the burial trenches are hydraulically connected. Fracture flow appears to take place in vertical joints and in the sandstone strata (O'Donnell, 1983).

Tritium, a radioactive isotope of hydrogen and the most abundant radioactive nuclide at the site, has migrated outside the restricted area, but apparently not yet off site. It is the only radionuclide detected in test wells, but it is not in the monitoring wells which ring the perimeter. Since transport from the trenches takes place in fractures, monitoring wells would have to be precisely located in fractures to detect radionuclide migration (Fischer, 1983). According to the US Geological Survey, tritium is migrating at the rate of 50 feet per year (O'Donnell, 1983). It has been detected in the sap of nearby maple trees, but it is not clear whether this is due to the evaporator or to subsurface in migration. Sampling of trees was done in March 1983: with the plastic cover in place the evaporator was shut down in December 1982.

Since 1978 the State of Kentucky has spent about $7m at Maxey Flats. Contracts with a geological firm and the purchase of the lease from NECO together cost about $2m. The State further estimates that it costs about $1m per year to maintain the site. Research and development costs are additionally paid by Federal agencies, particularly the NRC and DOE. Decommissioning plans have recently been formulated by the State of Kentucky (Kentucky, 1984). They will involve solidifying 80,000 gallons of evaporator sludges stored in the tank farm. To minimise further settling, the trenches will need to be compacted; then covered with asphalt, cement or vegetation to prevent water infiltration to the trenches. Lateral ground water flow will be eliminated with a cut-off trench, around the perimeter of the site. Excluding project management and contingencies, decommissioning costs range from $52m to $121m. These figures assume that the land reverts to unrestricted use after 100 years. If it indeed costs $121m to decommission the site, this is equivalent to $60 per cubic foot of waste material buried there. It is interesting to compare this real cost to the ideal long-term care and closure stabilisation costs of $2 per cubic foot quoted by a DOE contractor for a site of equivalent size (EGG, 1983b). Obviously DOE is anticipating that past problems will not be repeated at future radioactive landfills. In addition to these decommissioning costs, the State of Kentucky intends to buy a buffer zone around the site.

Sheffield

In 1966 permission was given to establish a low level radioactive waste disposal site near Sheffield, Illinois, in Bureau County, about 50 miles north of Peoria. The site is owned by the Department of Nuclear Safety on behalf of the State of Illinois, and, unlike the other commercial radioactive landfills in the US, it is directly licensed by the NRC. The burial of radioactive waste was first authorised in 1967. NECO, the same site operator as at Maxey Flats, but now called US Ecology, succeeded the original lessee in 1968 (NRC, 1981). The 20-acre site is in a region of abandoned coal mines, and is bordered to the north by rolling terrain and a 40-acre hazardous waste disposal site, also operated by NECO. The temperature in this midwest region varies between minus 10 degrees F. and 90 degrees F, with precipitation averaging 35 inches per year. The water table varies from 15 to 45 feet below the ground surface.

In ten years of operation three million cubic feet of waste containing 60,000 Ci of by-product material, 55kg of special nuclear material and 600,000lbs of source material were buried in 21 trenches (NRC, 1981). These wastes were dumped haphazardly to three feet of grade and the trench was then backfilled and contoured to aid water runoff. No detailed inventory of trench content exists. The trench caps are three feet thick. The last burial was in April 1978. The bottom of all trenches were supposed to be seven to ten feet above the water table, but one trench intercepts it.

In 1976, when space in the present 20-acre site was exhausted, NECO applied for a 168-acre expansion of the site, but the application was withdrawn at the recommendation of the State Attorney General. In 1979 NECO attempted to abandon Sheffield, and the site was officially closed. The NRC obtained a court injunction, forcing NECO to remain on the site. The perpetual maintenance fund for the site was exhausted, and the Attorney General entered a $97m suit against the operator.

The glacial deposits underlying the site have low permeability, but highly permeable pebbly sand underlies 60 per cent of the site and is the principal pathway for groundwater flow off the site near the east boundary (Johnson, 1983). Between 1975 and 1978 tritium moved 75 feet eastwards, but the migration rate has since accelerated. By 1982 tritium had migrated 665 feet from the closest burial trench and was discharging into a pond on private property. In 1983 NECO bought 120 acres as a buffer zone. According to NECO consultants movement velocities ranged between 0.4 and 2.7 feet per day. Recent chaser dye tests by the Illinois Geological Survey indicate a narrow tritium plume, now thought to be an underground stream, moving east at a rate of 2000 to 3000 feet per year (Chin, 1984). In the light of this, the State is reassessing whether to buy additional buffer acreage, and is also investigating the possibility that it should take over the site. Local citizens have requested that the radioactive waste materials should be exhumed and placed in above-ground storage.

The trench caps have severely eroded and cracked, and they require continual maintenance. Erosion occurs between the trenches, and cracking occurs at the side of the trench caps. Numerous incidences of subsidence, generally following heavy moisture, have also occurred. Most are small depressions about three feet in diameter, but some are 15 feet in diameter and 10 feet deep. Periods of prolonged dryness lead to trench cap cracking, and later erosion and water entering the trenches.

Since the trenches will settle differentially, compaction methods have been proposed. They consist of dynamic consolidation (dropping weights 5 to 40 tons from heights of 20 to 100 feet), pile driving (which would go through the waste containers and contaminate the equipment), surcharging (placing heavy loads on the trench caps) and blasting. Grouting the trench contents has also been proposed (NRC, 1981). The costs vary from $49,000 per trench (surcharging) to $854,000 per trench (grouting). It is not clear who would pay for these remedial measures.

West Valley

The West Valley site is 30 miles southeast of Buffalo, New York, in rural Cattaraugus County. The 3348-acre site was acquired by the State of New York in June 1961, and leased to Nuclear Fuel Services (NFS) two years later for constructing a nuclear fuel reprocessing plant (Resnikoff, 1977).

The site is on a plateau bounded on three sides by bedrock hills. Buttermilk Creek cuts through its centre at about 200 feet below the plateau. The substrate is highly impermeable silty till with up to ten feet of more permeable glacial till overburden. The silty till contains numerous vertical fractures and layers of more permeable sandy strata.

There are two different sites here. One is the burial site associated with construction of a reprocessing plant which was begun in 1963 and was started up in April 1966. During the six years operating life of the NFS plant 624 tons of irradiated fuel were processed. The high level wastes were placed in two underground tanks, and liquid and gaseous radioactive effluents were released to the Cattaraugus Creek watershed and atmosphere. NFS operated a seven-acre burial area, licenced by the NRC for solid radioactive wastes from plant operations: i.e. highly contaminated fuel hardware, obsolete equipment, and degraded process solvents 'absorbed onto a suitable solid medium' (DOE, 1978). Approximately 528,000 Ci of extremely radioactive materials (139,000 cubic feet) were buried in three by seven feet unlined holes, 50 feet deep. Half a ton of irradiated fuel elements was also buried in the NRC-licenced burial ground. Adjoining it is a commercial radioactive landfill licenced by New York State. This 25-acre site, seven acres of which have been used, is in the former Spitler's Swamp. Both sites have the same topographical features. Approximately 740,000 Ci of radioactivity in a volume of 2.3m cubic feet were buried in unlined trenches in seven acres of the landfill (Batelle, 1979). The commercial landfill ceased operation in 1975, when radioactive trench water began to seep through the trench caps. The landfill has two sets of trenches to the north and south. Within a few years of filling, water began infiltrating and accumulating in the north trenches. As at Maxey Flats, the cause of this was that the walls and bottom of the trench were more impermeable than the cap. Water rose in the trenches at an accelerating rate. In one case the water rose at one foot three inches per year from 1969 to 1971, at three feet three inches from 1971 to 1973 and at five feet per year from 1973 until the trench caps broke through in 1975. Each foot rise in the water level is approximately equal to 100,000 gallons of water. The radiation levels in this trench water reached up to 1,000 times the off-site maximum permissible concentration. The trench cap breakthrough actually exposed the waste materials. The covers were then repaired, and in 1978 an additional four foot cover was added, contoured and seeded. The water infiltration rates then dropped to one to 1.5 feet per year; the same rates as from 1969 to 1971. Water infiltration has not ceased, leading us to conclude that there may also be underground migration along sandy strata.

Based on the supposedly more successful experience with the south trenches, which are covered by an eight foot cap, Federal agencies were quite optimistic about the future performance of the north trenches. 'It is too soon to determine the effectiveness of these procedures, but experience with the southern trenches would indicate that infiltration through the caps should now cease and erosion should be prevented' (DOE, 1978). The performance of the south trenches, which began accepting waste in 1969, was initially better. The water levels remained constant from 1973 to 1977. Then from 1977 to 1979, at the same time that DOE was extolling the virtues of the eight foot covered south trenches, the water levels began to rise at a rate of two feet two inches per year (Cashman, 1982). The dry summer of 1979 may have caused the clay trench covers to crack more than usual, thereby providing a water infiltration route for the following winter and spring. The infiltration rate increased in 1980 and the State began pumping the 'improved' south trenches to avoid trench cap breakthrough. The pumping operation - a process called 'controlled leakage' - involves transferring the radioactively contaminated water to a treatment facility, where the water is passed through ion-exchange resins and then released to the Cattaraugus Creek watershed. Obviously this treatment method releases tritium. The radioactive sludges are then buried in the NRC-licenced burial area. As at Maxey Flats, the West Valley trenches contain high levels of the chelating agents EDTA and DPTA (NRC, 1984a). (A chelating agent is a solvent used to decontaminate piping in nuclear reactors by making cobalt-60 and other radionuclides soluble.) It is likely that the presence of these chelating agents leads to the complexing of cobalt, plutonium, americium and strontium (complexing is the chemical combination of hydrocarbons with, in this case, radionuclides, which increases the water solubility of the latter). The treatment facility has questionable effectiveness in the presence of these agents.

Another serious concern at both burial areas is rapid erosion of the plateau banks towards the burial ground area (NRC, 1984b). In 1978 the steeper trench slopes were covered with a plastic sheet and a layer of crushed stone 'to anchor and protect them from sunlight' (NRC, 1980). This is clearly only a temporary solution.

In order to attempt to eliminate continual maintenance at the West Valley site, the State of New York State Energy Research and Development Authority, has contracted with DOE to study how to decommission the burial areas. While much attention was focused on the State-licenced commercial burial ground, little notice was paid to the NRC-licenced burial area. The presence of high tritium levels in the ravine between the State-licenced and NRC-licenced burial areas led an official of the New York Department of Health, in rare agreement with citizen groups, to call for exhumation of the NRC-licenced area (NYSDOH, 1979). Still, any potential problems with the NRC-licenced area would have gone unnoticed had the New York State Geological Survey not struck oil on the north slopes, or, more properly, struck kerosene contaminated with plutonium. It was detected twenty feet below ground, approximately 60 feet from the nearest potential source. The radioactive concentrations were about 1,000 times the off-site maximum permissible concentrations, due primarily to iodine-129. Since the Federal agencies and New York State Geological Survey had assured everyone that the silty till was impermeable, this plutonium migration was surprising.

After irradiated fuel is dissolved in nitric acid, the organic substance Tributyl-n-phosphate (TBP) is added, and it selectively removes uranium, plutonium and some fission products. Eventually this solvent becomes too contaminated and must be disposed of. At West Valley it was buried in 1,000 gallon tanks in the NRC-licenced burial area; in the case of Windscale it was disposed of in the Irish Sea. However, it is now believed that kerosene with TBP increases the permeability of silty till, and a radiation plume, over 30 feet wide and of uncertain depth, is now moving towards a feeder stream of Buttermilk Creek, Erdman's Brook, to the north of the NRC-licenced burial area. A string of interceptor wells has been placed at the edge of the plateau, near Erdman's Brook. The NRC is funding a major research and development programme to determine how to stem the tide. At present, wells have been drilled near the likely leaking tanks and radioactive solvent is being removed as it accumulates. An added complication is the fact the kerosene floats on water, thereby rising and falling seasonally. Fortunately, the likely kerosene sources contain very little radioactivity, less than one Ci total. However, carbon steel tanks in nearby holes contain about 1100 Ci in organic solvent. Since they will eventually degrade, environmentalists have called for them to be exhumed before plutonium begins to migrate.

Barnwell

The 280-acre Barnwell landfill is the major operating landfill for radioactive waste in the US. In 1982 approximately 35,000 cubic metres, or 46 per cent of the radioactive waste volume, and about two-thirds of the radioactivity, were buried at Barnwell (EGG, 1983a). In 1979 the Barnwell site accepted two-thirds of the nation's waste volume (NUS Corporation, 1980) until the Governor of South Carolina imposed a monthly limit of 100,000 cubic feet. Much of the excess now goes to the Richland landfill. The Barnwell site is owned by the State of South Carolina, which has licenced Chem-Nuclear Systems, a subsidiary of Waste Management Inc., to operate it until 1992. The State has made clear its preference for closing the site at that time, when supposedly another state in the southeast region will open a facility. To equalise the waste burden, Congress enacted the Low Level Radioactive Waste Act of 1980, which requires each region of the country to select and operate waste facilities (see Chapter 7 of this book).

The Barnwell site is about 70 miles southwest of Columbia near the Savannah River. Originally licenced in 1972, it has a large capacity; 2.4m cubic metres. Unlike Maxey Flats and West Valley, it is in permeable sand and clay. Water entering the trench caps does not accumulate, but passes directly through the bottom of the trench. No trench water samples have yet been collected, even though the mean annual precipitation is over 43 inches per year. Thus, unlike Maxey Flats and West Valley landfills, radioactive materials do not 'brew' in trench water here. Surface water runoff is also minimal because of the sandy soil and flat topography. Compared to the northern sites, little radioactivity has migrated out of the Barnwell landfill.

Radioactive waste is buried with the standard cut and fill burial method used at other sites. In the early 1970s waste was buried quite haphazardly. Barrels were often rolled out of the rear of lorries, in what is called the 'kick and roll method'.

At present all waste materials must be solid, and they are packed tightly within the trench in order to minimise voids. The problems of package degradation, trench cover subsidence and water infiltration are no different at Barnwell than at the three closed sites in the north, but little radioactivity has migrated because of the most permeable nature of the surrounding medium.

Some tritium migration has occurred. In 1981 it was detected 21 metres deep, indicating downward movement of water from the trenches. Above-ground levels of tritium have also been detected as far as 75 metres southwest of buried waste (Cahill, 1982). The only other radionuclide to have migrated was cobalt-60. Since the same chelating agent that migrated at the NRC-licenced burial ground at West Valley, TBP, was buried at Barnwell until 1982, it is suspected that the cobalt-60 and strontium-90 levels will begin to migrate as the waste packages degrade.

Richland/Beatty

The Richland, Washington and Beatty, Nevada sites are, respectively, in semi-arid and arid parts of the US. The absence of precipitation greatly simplifies waste management procedures. The trenches remain open until completely full. Unlike those in the east, no attention is paid to eliminating voids, and the trench caps are not compacted. No sampling of trench water needs to be carried out. The one serious disadvantage of these sites is that they are far from major waste generators in the east, thereby greatly increasing transport costs and impacts. Wind erosion and animal movement (gophers burrowing into trenches, for example), are slight problems at these western sites, but they scarcely merit attention.

The 90-acre Beatty site is near Death Valley, 112 miles northwest of Las Vegas. The annual precipitation is only four inches. The soil is sand and gravel. The only problems involved radioactive drums that were buried outside the site boundary, and radioactive materials that were removed from the site (and later returned). In 1982 Beatty accepted two per cent by volume of the nation's waste (EGG, 1983a).

The 110-acre Richland site is near the Oregon border, at the confluence of the Yakima and Columbia Rivers. The annual precipitation is eight inches. The soil is also sand and gravel. In 1982, Richland accepted 52 per cent of the nation's waste, becoming the major landfill for the US.

If present decontamination practices are maintained, it is likely that these two dry sites will receive large volumes of wastes from decommissioned reactors. These wastes contain chelating agents that react poorly when mixed with wastes containing toluene and xylene (NRC, 1983). Solidified decommissioning wastes with chelating agents showed up to a 35 per cent weight loss when immersed in representative trench liquids. At present these decommissioning wastes must be disposed of in a dry environment. Unless the NRC forces a change in the waste management practices at reactors, by the end of the 1990s much waste from decommissioned reactors will go to the west. By requiring wastes to go from one region of the country to another, the NRC is overriding the Low Level Radioactive Waste Act of 1980, which requires each region of the country to handle its own waste.

WHAT HAS BEEN LEARNED?

Clearly, radioactive landfills in the northern humid regions have operated poorly, and those in the arid regions have operated satisfactorily. Even in the former the off-site radioactivity concentrations in water have been low. While trench water itself has been about 1,000 times maximum permissible concentrations, dilution has greatly reduced the concentrations while generally increasing the number of people contacted. But calculations have not fully taken account of the full dose to present and future populations. All would agree that it must be the aim of regulatory agencies to minimise radiation exposures and greatly improve past performance. Rather than stabilised, maintenance-free landfills, we have sites that have required active maintenance within ten years of trench closure.

The problems at the three landfills in the humid north have been due to:

1. Excessive groundwater. This has partly arisen from the lack of siting criteria, and licence and information requirements at the time. Contrary to the NRC view that the 'final siting decisions have been based largely on hydrogeologic and economic factors' (NRC, 1980), according to the Illinois Geological Survey 'the geologic descriptions were fairly superficial' (IGS, 1982). Hydrogeology was only a factor if it was the intent to site one landfill in a swamp (West Valley) and another in an underground stream bed (Sheffield);

2. Degradation of waste containers and the waste form itself;

3. Subsidence, cracking and erosion of trench caps, which has led to water infiltration into trenches - called the 'bathtub effect';

4. The presence of chelating agents, which have rendered soluble the normally insoluble elements, americium, caesium, strontium, plutonium and cobalt;

5. Lack of stabilisation and long-term care funds, which has passed the financial burden from the waste generators and disposal companies to the State and Federal taxpayer.

The public perception of both radioactive and toxic chemical landfills is that they do not work in the humid north. Landfills are like tea bags: the water comes in; the flavour goes out. It is clear that there will be intense public opposition whereever a future radioactive landfill is proposed. It would be best to proceed as if landfills were not an option. Clearly some uses of radioactive materials are absolutely essential to a modern society, until safer substitutes are devised. To care for this minimum amount of radioactive materials, long and short lived materials should be allowed to decay to non-hazardous levels. Voluminous materials should be super-compacted. This waste could be stored in above-ground storage facilities with leachate collection systems, as is done in many parts of the world. The longer-lived and more hazardous low level waste should be reclassified as high level waste, and should be disposed of deep underground. More generally, as a matter of equity, we should not transfer the waste burden geographically or over time. Those who use radioactive materials which produce waste should pay the cost. As a society, and from general principle, we should generate only those necessary hazardous materials we can safely manage.

REFERENCES

DOE (Department of Energy) (1978) Western New York Nuclear Service Center Study, Companion Report, TID-28905-2, Washington DC.

EGG (EG & G Idaho, Inc.) (1983a) 1982 State-by-State Assessment of Low-Level Radioactive Wastes Shipped to Commercial Disposal Sites, DOE/LLW-27T, Idaho Falls.

EGG (1983b) An Analysis of Low-Level Waste Disposal Facility and Transportation Costs, DOE/LLW-6Td, Idaho Falls.

FISCHER, J. N. (1983) 'US Geological Survey studies of commercial low-level radioactive waste disposal sites - a summary of the results', Proceedings of the Fifth Annual Participants' Information Meeting: DOE Low-Level Waste Management Program, EGG Idaho, Idaho Falls.

HEALEY, R. W. (1983) 'Preliminary results of a study of the unsaturated zone at the low level radioactive waste disposal site near Sheffield Illinois', Proceedings of the Fifth Annual Participants' Information Meeting: DOE Low-Level Waste Management Program, EGG Idaho, Idaho Falls.

IGS (Illinois State Geological Survey) (1982) 'A geological case history: lessons learned at Sheffield Illinois', Champaign, Illinois.

JOHNSON, T. M. (1983) 'A study of trench covers to limit infiltration at waste disposal sites', Proceedings of the Fifth Annual Participants' Information Meeting: DOE Low-Level Waste Management Program, EGG Idaho, Idaho Falls.

KENTUCKY (1984) Comprehensive Low-Level Radioactive Waste Management Plan for the Commonwealth of Kentucky, Kentucky Natural Resources and Environmental Protection Cabinet, DOE/ID/12348-T6, Frankfort, Ky.

MEYER, G. L. and BERGER, P. S. (1975) 'Preliminary

data on the occurrence of transuranium nuclides in the environment at the radioactive waste burial site, Maxey Flats, Kentucky', International Symposium on Transuranium Nuclides in the Environment, IAEA, San Francisco.

MILLS, D. and RAZOR, J. (1983) 'An infiltration barrier demonstration at Maxey Flats, Kentucky', Proceedings of the Fifth Annual Participants' Information Meeting: DOE Low-Level Waste Management Program, EGG Idaho, Idaho Falls.

NRC (Nuclear Regulatory Commission) (1980) Technology, Safety and Costs of Decommissioning a Reference Low-Level Waste Burial Ground, NUREG/CR-0570, Washington DC.

NRC (1981) Evaluation of Trench Subsidence and Stabilisation at Sheffield Low-Level Radioactive Waste Disposal Facility, NUREG/CR-2101, Washington DC.

NRC (1983) Physical Tests on Solidified Decontamination Wastes from Dresden Unit 1, NUREG/CR-3165, Washington DC.

NRC (1984a) Scoping Study of the Alternatives for Managing Wastes Containing Chelating Decontamination Chemicals, NUREG/CR-2721, Washington DC.

NRC (1984b) Geologic and Hydrologic Research at the Western New York Nuclear Service Center, West Valley, New York, NUREG/CR-3782.

NUS (NUS Corporation) (1980) The 1979 State-by-State Assessment of Low-Level Radioactive Wastes Shipped to Commercial Burial Grounds, NUS-3440, San Francisco.

NYSDOH (New York State Department of Health) (1979) News release, Albany, NY.

O'DONNELL, E. (1983) 'Insights gained from NRC research investigations at the Maxey Flats LLW SLB Facility', Proceedings of the Fifth Annual Participants' Information Meeting: DOE Low-Level Waste Management Program, EGG Idaho, Idaho Falls.

ORNL (Oak Ridge National Laboratory) (1984) Plan for Diagnosing the Solvent Contamination at the West Valley Facility Disposal Area, ORNL-2416.

RESNIKOFF, M. (1977) 'Sierra Club testimony related to Section IV E, Reprocessing, Final GESMO I, March 4', NRC Docket No. RM-50-5, Washington DC.

SIERRA CLUB (1984) 'Low-level nuclear waste: options for storage', Buffalo, NY: Sierra Club Radioactive Waste Campaign.

Chapter Nine

DECOMMISSIONING AS A NEGLECTED ELEMENT IN NUCLEAR POWER PLANT SITING POLICY IN THE US AND UK

M. J. Pasqualetti

Nuclear siting discussions have almost always emphasised reactor operation, concentrating on locational influences like seismic stability, availability of cooling water, and future markets. Siting questions associated with post-operation, or 'back-end', phases of the fuel cycle such as waste disposal and decommissioning were slower to attract attention. The effects of this oversight, in the form of waste disposal problems, emerged a few years ago and continue to plague the industry (see this volume Chapter 6). But decommissioning is still a vague topic to the private citizen and has been only broadly considered by most electricity undertakings. This situation is just now beginning to change. In anticipation of the first wave of decommissioning in the US, the Nuclear Regulatory Commission has published a proposed rule on related long-range policy issues (US NRC, 1985). In the UK the first glimmer of public interest in decommissioning occurred at the official inquiry into the proposed construction of Britain's first pressurised water reactor (PWR) at Sizewell, on the Suffolk coast (see this volume Chapter 4; O'Riordan, 1984). In both countries an increasingly sophisticated approach to decommissioning will undoubtedly produce related, and interdependent, impacts on siting and siting policy.

The literature on siting, especially power plant siting, is enormous and intrinsically geographical, as is reflected in the long record of publication on the topic by professional geographers (Pasqualetti, 1985). It is through a geographical orientation that many unrealised relationships between siting and decommissioning may be identified. This interest in siting has recently been reinvigorated, both in the US (e. g. Dobson,

1984; Pasqualetti, 1985; Richetto, 1984) and the UK (e.g. Openshaw, 1980, 1982a; Openshaw and Taylor, 1981). Partly this renewed attention was a reaction to stiffer environmental requirements which were being made of power utilities, and partly it was a response to the emergency planning issues raised by the Three Mile Island accident (Hansis, 1980; Openshaw, 1982b, 1986; Zeigler and Johnson, this volume Chapter 12).

As utility companies have found it increasingly difficult to find acceptable sites for their power plants, alternative siting schemes have been forwarded (e.g. Baker, et al., 1980; Solomon and Haynes, 1982; Solomon, Haynes, and Krmenec, 1980). Neither these alternative schemes nor the more conventional approaches have considered decommissioning as a siting factor. This is understandable in view of the low level of public interest in and knowledge on the topic in general. But as a result, the substantial potential (generally or in terms of power plant siting) which decommissioning has to destabilise already shaky nuclear programmes has been skipped over.

This chapter examines decommissioning evidence presented at the Sizewell 'B' Public Inquiry and combines it with more generalised examples from the United States. The purpose is to assess the importance of decommissioning as a factor in power plant siting by examining how siting and decommissioning decisions affect one another geographically. This assessment demonstrates that because the discussion of decommissioning has heretofore been fundamentally technical in orientation and motivation, it has not considered significant relationships which are as basic as those that concern facility siting.

The examples from Sizewell and from the US indicate that:

1. siting decisions affect decommissioning options;

2. decommissioning options, in turn, affect land use planning;

3. topographic and vegetative camouflage at the site affect public opinion and thus decommissioning options;

4. aesthetic qualities at a site can justify and perhaps precipitate certain decommissioning actions;

5. the type and sequencing of decommissioning can influence the length of the land commitment;

6. policy decisions which the utility company makes about extending plant and site life, the type of decommissioning, and so on, are all factors significant to land use in general and the site in particular.

The chapter is divided into two basic parts. It begins by summarising the essentials of decommissioning by reference to experience in the United States and the United Kingdom. The second part is made up of a discussion of the links between land use and decommissioning mentioned above. The discussion illustrates spatial influences on nuclear development which have not yet been fully examined. It also illustrates the range of implications which nuclear decisions can have; a range of such breadth and complexity as to tempt official sidesteps rather than direct and thorough appraisal. The recent awakening interest in decommissioning also illustrates the incremental nature of nuclear power policy formation which characterises and reflects the inherent complexity and longevity of nuclear programmes.

Nuclear decommissioning

Everything in the nuclear industry is built with a certain life expectancy. During this life, all products pass through several distinct phases: design, construction, operation, retirement. Each product is financed with this life expectancy in mind. For nuclear generating stations the term 'decommissioning' has been given to the retirement phase. This is the time when continued operation is uneconomic or unsafe. Much like an automobile, nuclear power plants are expected to operate most efficiently and cost-effectively during early life. Unlike most automobiles, nuclear facilities receive a constant and programmed amount of careful maintenance. Even with such attention, however, the nuclear utility company, like the car owner, must contend with gradual deterioration, with changes in standards and laws, and with the constant availability of technically superior components. Collectively, all these factors result in diminishing returns and eventually the decision for abandonment.

The main technical reason for decommissioning a nuclear plant is the loss in ductility (pliability) of the reactor vessel, the result of prolonged neutron bombardment leading to embrittlement of the vessel. There are several possible remedial actions when this stage is reached, including replacement of the reactor vessel. Ultimately a point is reached when further repairs are not the most viable economic choice, especially as alternative and superior technologies may have become available. This point is expected to occur after 35-40 years, but it may be sooner or later.

It is the radiological hazard that accounts for the particular attention which decommissioning in the nuclear industry is beginning to receive. Of course, all private, governmental, and commercial nuclear facilities (e.g. uranium mines and mills, uranium conversion and enrichment plants, fuel fabrication and reprocessing structures) must be decommissioned. Currently, the commercial nuclear power plant attracts the greatest public interest.

The three basic types of decommissioning are immediate dismantling, safe storage, and entombment (Fieldman, _et al_., 1981). All have been used in the past, but only on small non-commercial reactors; no large or fully commercial reactor has yet been decommissioned. Immediate dismantling (sometimes called DECON by the nuclear industry) involves decontaminating and promptly removing all radioactive materials to allow release of the site for unrestricted use. It may also involve removing the containment structure, foundations, roads, powerlines, and the like. Safe storage (SAFSTOR) is a procedure designed to allow radioactive decay of some of the dangerous but shorter-lived radioisotopes, such as Cobalt-60. It involves removing the spent fuel and securing and maintaining the property so that the level of risk is acceptable for an extended decay period (perhaps up to 100 years). After some additional actions such as removing any still-hazardous materials, the facility itself can be released for unrestricted use or it can be dismantled. Under the entombment (ENTOMB) option the radioactive or contaminated materials are shielded in concrete rather than removed. This is a viable option only if the entombing structure can be expected to last for several half lives of the most objectionable and long-lived isotope.

For PWRs such as that envisaged at Sizewell, the decommissioning sequence would normally consist of three stages. Stage one entails removing fuel from the reactor and putting it in a fuel storage pond (for five years), draining the reactor system and boron recycling tanks, and decontaminating the reactor system. Stage two involves removing both radioactive and non-radioactive parts of the plant, including turbine-generators, main electric plant, the turbine house, reactor main systems with the reactor building, nuclear auxiliary systems in auxiliary buildings, and radioactive waste systems which are located in a separate annex. Stage three involves removing the insides of reactors - the reactor pressure vessel and insulation, and the biological shield and cavity liner. As suggested above, the third stage could be done early or deferred (Gregory, 1982a).

The type, timing, and sequencing decisions which are made for each power plant will have impacts and influences which vary from one plant to another depending upon site characteristics and geographical relationships within the local area and on the nuclear fuel cycle itself. In the present study several power plants in the US and the UK were examined.

Power plant and site characteristics in the United States

There are about 70 commercial nuclear power plants in the US, concentrated largely east of the Mississippi River. In a survey for this study, attention was focused either on those considered 'eligible' for decommissioning in the near future or on those which have already been decommissioned (Figure 9.1). Of the former, attention was concentrated on those plants which had been inoperative (but not decommissioned) for an extended period or had attracted interest because of chronic operational or structural difficulties (e.g. pressurised thermal shock) (Dircks, 1982). The usual bibliographic sources were supplemented by a series of telephone interviews with representatives of the operating utilities, during which questions were asked about possible post-operational use of the land, existing and potential land use conflicts, proximity to residential areas, plans for future site use, expected mode of decommissioning, State and local decommissioning requirements, and extant decommis-

Figure 9.1 SITES INVESTIGATED IN THIS STUDY

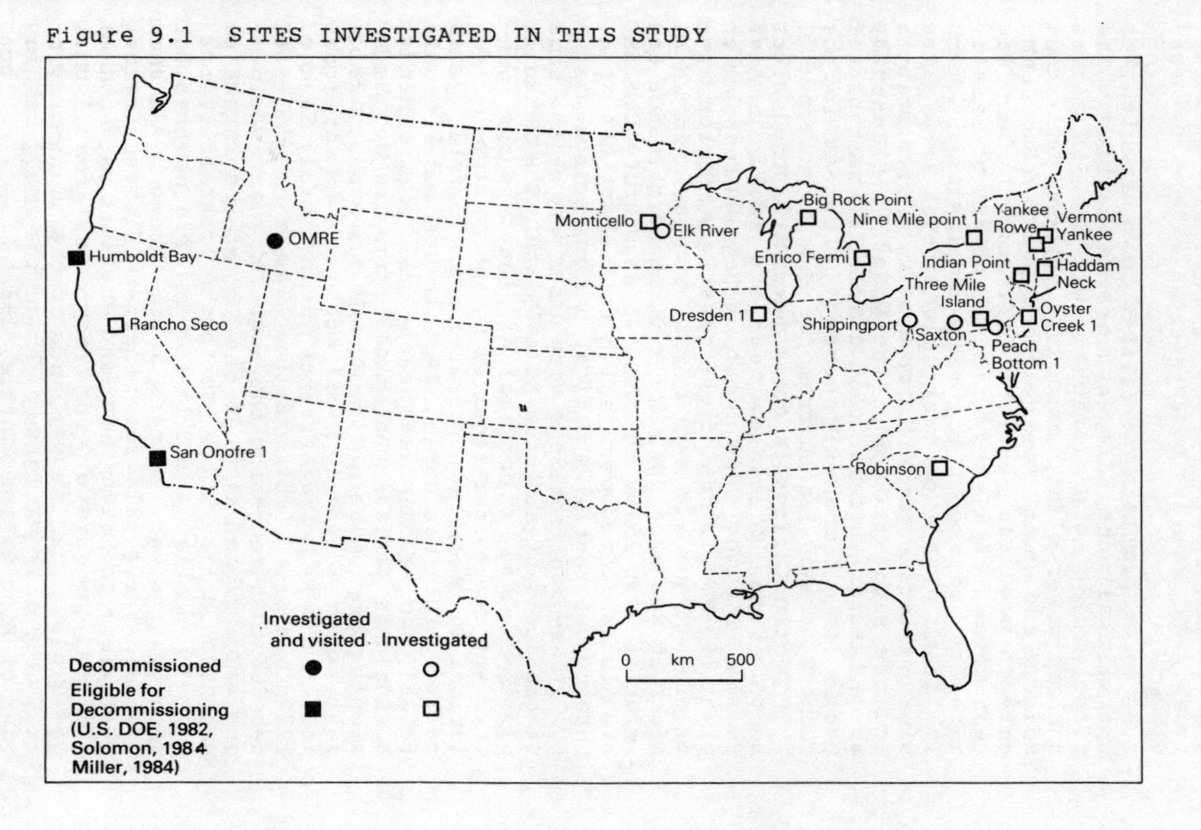

sioning documents. Regulatory agencies, private consultants, public interest groups, private individuals, and lobbying organisations were also interviewed.

Three plants were visited, one of which was operational, one idle, and one dismantled. The first, which is operational, was the San Onofre nuclear power plant, owned by Southern California Edison, and on the southern boundary of San Clemente in populous Southern California. Unit One (436 megawatts (MW)) of this plant was of primary interest because of its age (commissioned 1968). It, along with two newer units, is on an 87-acre site cut into a marine terrace between Interstate-5 and the Pacific Ocean. The plant is flanked by State beaches which attract crowds for most of the year. It is the only commercial plant in the country on land leased from the Federal government.

On the California coast several hundred miles north of San Clemente, near the small city of Eureka, the now idle Humboldt Bay nuclear power plant (63 MW) is one of the first-generation commercial US nuclear facilities (commissioned 1963). After it was shut down for routine maintenance and refuelling in 1976, the Nuclear Regulatory Commission (NRC) denied permission to restart the plant pending substantial modifications for seismic safety. These modifications were under way at the time of the accident at Three Mile Island. The Humboldt plant, like all other nuclear plants in the country, was thereafter required to adhere to new regulations on emergency preparedness. Several years passed while the owner, Pacific Gas and Electric (PG&E), worked to meet these new requirements. During this period, public calls for its decommissioning grew more noticeable (e.g., Rubin, 1981), particularly from a local public interest group called the Redwood Alliance (Nader, 1981; Savage, 1979; Zichella, 1983). By the summer of 1983, PG&E decided that reopening the Humboldt plant would be uneconomic, and it told the NRC of its intent to decommission the unit. It is the first fully commercial nuclear plant in the world announced for decommissioning. Members of the Redwood Alliance continue their call for immediate dismantling but PG&E plans to leave the plant where it is, in safe storage, immediately adjacent to two fossil fuel power plants in the wetlands to the south of Eureka, near busy US Highway 101 and the small fishing town of King Salmon (Tuttle, 1982).

The third visit was to the Federal Idaho National Engineering Laboratory. The laboratory consists of about 2315 sq km (926 sq miles) of sagebrush and basalt-covered land on the Snake River Plain in Southeastern Idaho at an average elevation of 1483 m (4863 ft.). The first nuclear reactor to be built in the US and the first reactor to generate electricity, as well as 40 other governmental reactors of great variety, have all dotted this plain. Most of these small research reactors have been removed from service; many of them by dismantling (Clark, 1983). One reactor, the Organic Moderated Reactor Experiment (OMRE), was dismantled completely - including all foundations, contaminated dirt, powerlines, and signs. The debris was placed in waste repositories elsewhere at the laboratory (Rasmussen, 1977; Hine, 1980). OMRE is the best example of complete dismantling and site reconstitution in the US and probably in the world.

The information on land use and power plant site selection which was gained during these visits and interviews was assessed qualitatively. It has been added to the pre-construction appraisal of a single plant in the United Kingdom.

The United Kingdom

As in the US, older reactors in the UK are nearing a period when procedures and decisions about their decommissioning must be established. For several reasons, the wide-ranging Sizewell 'B' Public Inquiry and the mood that was found there (see this volume Chapter 4), presented a unique opportunity to examine the decommissioning of nuclear generating stations. First, it was an open forum on many topics which were linked directly with decommissioning (e.g. waste disposal). Second, professional and technical ties to other countries (e.g. fuel from Australia, reactor design from the US) meant that the evidence presented had applicability elsewhere. Third, the Inquiry allowed the first opportunity for a dialogue (albeit indirect, limited, and controlled) between industry and the public on the topic of decommissioning policy. Fourth, all documents submitted to the Inquiry became available publicly.

Siting criteria for Sizewell are basically the same as in the US. In his proof of evidence on site selection, Mr Gammon, of the Central Electricity Generating Board (CEGB) enumerated the main requirements for a power station site:

a) a location strategically placed for connection to the transmission grid and so positioned as to supply electricity to areas with large demands;

b) an adequate area of reasonably flat land at an acceptable height above the source of cooling water;

c) foundation conditions capable of supporting heavy loads under all conditions;

d) an abundant unfailing supply of water for condenser cooling;

e) acceptable environmental effects. (Gammon, 1982, p. 2).

The site for Sizewell 'B' unit is alongside the Magnox Sizewell 'A' reactor on the Suffolk Coast 35km (22 miles) northeast of Ipswich and 31km (19 miles) south of Lowestoft. The Sizewell site was once heathland and conifer plantations, with low grade agricultural land and marshland to the north (Gammon, 1982, p. 6). It is an area of very low relief and, especially along the coastline, unusually long vistas.

Links between siting and decommissioning

Decommissioning can be broadly divided into technical and non-technical considerations. Technical aspects include such matters as ductility, cutting techniques, and radiation monitoring. For convenience all matters not obviously technical may be categorised as 'non-technical', even though one cannot separate the two categories so completely. Throughout the decommissioning literature the obvious emphasis has been on the technical side of the matter. Although there has been some recent non-technical attention directed toward a range of financing schemes, the general thoroughness of research on the development and safety issues of nuclear power has yet to be applied to the topic of decommissioning.

One way to illustrate the industry's current predilection for technical studies is to focus on a sub-set of non-technical topics. The sub-set used here is the link between decommissioning and the power plant site. The goal is to determine whether the links between decommissioning and siting have been identified and adequately addressed. If they have not, one could assume that other relationships might not have been identified or been given sufficient consideration either. The decommissioning /siting matter is discussed below in terms of topography, visibility, and aesthetics; land commitments and unrestricted use limits; waste disposal sites and transport routes, planned land use; and single-versus multiple-unit sites.

Topography, visibility, and aesthetics

The degree of public opposition to nuclear power is partly perceptual, a function of visual awareness of physical facilities and their aesthetic intrusion. Visibility, in turn, is related to topography, vegetation and other screening materials, proximity to people, and frequency of passers-by. The Humboldt Bay and Indian Point power plants are both regularly and easily visible to many people (for location, see Figure 9.1). Compared to public reaction to other power plants, they have also long attracted a disproportionate amount of attention from those who wish them decommissioned. Other power plants such as San Onofre and Haddam Neck are also in urban areas, and both have provoked frequent calls for decommissioning. Many nuclear power plants are on coasts, and when these coasts are used heavily for recreation, the aesthetic argument for decommissioning becomes more persuasive. This has been experienced at San Onofre, which is between two attractive State-run beaches.

Coasts near Humboldt and Sizewell are used lightly for recreation, but both are noted for their natural qualities. The concern over the Sizewell site is partly the result of its location near or in several designated areas of high aesthetic value. For example, the site is within '... the Suffolk Coast and Heaths Area of Outstanding Natural Beauty statutorily designated in 1969 and is part of the Suffolk Heritage Coast identified in 1970 and completely defined in 1979 ... The Royal Society for the Protection of Birds Reserve at Minsmere lies 2km (1 mile) to the north of the site. The Nature Conservancy Council's National Nature Reserve at Westleton Heath is 5km (3 miles) to the north, and the Sites of Special Scientific Interest at North Warren some 3km (2 miles) to the south and at Minsmere Level immediately to the north ...' (Gammon, 1982, p. 7). The sequence, timing, and planning for decommissioning will have a direct bearing on how long the intrusion on these areas will continue.

The extraordinarily flat terrain at Sizewell (where the 'A' station is already built) will increase the aesthetic intrusion of the 'B' station's structures and this factor will be significant when the time comes to choose the type of decommissioning. As the Royal Institute of British Architects (RIBA) stated at the Inquiry: 'Some [buildings] (and in our view Sizewell 'B' is such a building) would remain visually hostile and unwelcoming' (RIBA, n. d., para. 1.14). As long as it remains on the coast, this large building (70 metres (230 feet) high and 55 metres (182 feet) in diameter, will occupy an area 'with virtually no other building development ... not even a coast road' (RIBA, n. d., para. 1.2). The decommissioning option of dismantling has the obvious advantage of early removal of the intrusions, and this appears desirable.

Thus, one may argue that choices about decommissioning options will be influenced, at least in part, by factors of topography, visibility and aesthetics, because these factors play a role in public awareness and subsequent public involvement in the decision making process.

Land commitment

The type and timing of decommissioning (along with whether it can actually be carried out safely and successfully) are such basic uncertainties at present that they will influence any discussion about associated land commitments. In the absence of commercial decommissioning experience, it is risky to predict its effects on socio-economics and land use planning. Nor is it possible to judge the timing or size of waste repositories, estimate the likely level of public pressure to re-use the power plant sites or extend the life of the plant itself, or gauge when naturally attractive areas will be aesthetically restored. In both the US and the UK at present it is possible that dismantling could be deferred for about 100 years. Such deferrals increase the uncertainties and make the job of the land use planner substantially more difficult.

Sizewell has received much attention because of its isolated and undeveloped location. This takes on a great significance because the period of the intrusion depends directly upon the type of decommissioning planned and the actual decommissioning plan itself.

The length of time that land will be committed at the Sizewell plant will be a function of CEGB policy towards the continued use of existing power plant sites, on whether the plant is put into prolonged safe-storage, and on the extent to which radioactivity can be removed from the site after plant dismantling. In general it is the CEGB's practice to continue developing existing sites 'where possible and practicable, if they satisfy electricity supply system requirements' (Gammon, 1982, p. 3). Sizewell 'A' (along with its related transport and transmission network) has been operating on the site for many years, and this historical land use was integral to siting Sizewell 'B' at the same place. Further units are also possible on this site.

There is no commitment by the CEGB to a particular type of decommissioning, although it is generally agreed that delayed dismantling would be employed (Gregory, 1982b). Thus, final dismantling would not occur until approximately the year 2139. Although, as Blake (1985) said, 'there is no commitment to this course of action', further delay would not appreciably reduce the radioactive hazard from the long-lived isotopes, such as those of nickel-59 and niobium-94.

One of the factors influencing the decision about land commitments and the type and timing of decommissioning will be the occupational radiation dosage which would accrue to those involved in the actual operation and to those who might be in the path of dust deposited from the project. Longer periods for radioactive decay would reduce worry about such dosage.

Once it is cleared (depending of course, on any decision about an even newer plant) there is no assurance that the land can be made suitable for unrestricted use. Alan Gregory, head of CEGB's decommissioning programme, said that 'The effects of levels [of radioactivity needed to be achieved for unrestricted use of the land] ... are still being established and there is reluctance to set a figure now until one is sure that the figure which the regulatory authority sets is going to be properly balanced between the safety levels and the reality of achieving it' (Gregory, 1984, p. 44). In other testimony, Mr Gregory illustrated the uncertainty over unrestricted use limits when he said that while it would probably be alright for people to walk over the land, there might be problems if it were to be occupied and used for farming (Gregory, 1982b, pp. 23-24).

The NRC and the CEGB intend to release as many nuclear sites as possible for unrestricted use. Once the required decontamination is complete, the choice affecting future use would be of no further official concern to the regulatory authorities unless a new nuclear facility were constructed. At such a time, the facility and its final disposition would become the responsibility of the state or county and various non-federal local agencies in the US. In the UK responsibility would lie with the State and its agencies. In the US, if funds were unavailable for dismantling the decontaminated structure could conceivably remain at the site indefinitely, regardless of what other types of land use were encroaching upon it.

Noting these uncertainties, the Royal Institute of British Architects said that a 'decision for consent to the construction of Sizewell 'B' is a decision to remove this area of land from human presence forever' (RIBA, n. d., para. 1.43). The Town and Country Planning Association also expressed misgivings about the length of the commitment when its representative said that the commitment of the land for the 35 years of plant operation and a further 110 years for radioactive decay 'represents a major environmental disbenefit' (Blake, 1985, p. 26).

The commitment of land to nuclear power plants must be considered both in terms of amount and duration. With regard to the former, the issue is the amount of the surrounding land actually in view of the power plant, rather than the land physically occupied by the power plant and its access routes. With regard to the duration of disruption, other factors such as the unrestricted use limit add a complicating factor.

Waste sites and transport routes

The existence, location, and distance of waste sites will have a strong bearing on the type of decommissioning possible. There is no commercially available waste facility for high level waste in the US or the UK; this is a well recognised problem in disposing of the waste produced through the lifetime operation of every power plant and it will be amplified when power plants are decommissioned. Additionally, there is insufficient low level storage to accommodate many decommissioned plants (Feldman, 1983; Resnikoff, this volume Chapter 8). Until ample and proper waste repositories are available, all decommissioned commercial plants will remain on the landscape in safe storage, along with any associated spent fuel (US Comptroller General, 1980). This situation eliminates the option of civilian re-use of nuclear sites, and it also leads to an accumulation of nuclear facilities on the landscape, regardless of the wishes of regulatory agencies, or neighbours.

Plant sites will be linked to waste sites by transport routes of an unspecified pattern and length. One of the intentions of the US low-level regional waste compacts (i. e. arrangements among groups of states - see this volume, Chapter 7) is to limit these transport distances. Transport routes to the high-level repositories, however, will be longer because there are fewer of them. Distances between plants and repositories will be shorter in the UK although the generally higher population density will result in more trips through inhabited areas. Inasmuch as none of the facilities is yet in operation, it is impossible to determine what routes the waste will travel, or what the potential might be to restrict legally such transport. Obviously, the locations of power plants and waste repositories will strongly affect the route selection and the use of land along these routes.

When repositories do not yet exist, it adds a layer of uncertainty about the routes to them, or about whether or not the wastes will have to be retained at the site of the power plant itself (Solomon and Cameron, 1984). Mr Passant (head of the Active Waste Project in the Generation Development and Construction Division of the CEGB) said that because of the long term periods expected before dismantling, 'the disposal options and sites available, and hence the waste acceptance criteria, are likely to be different from that at present' (Passant, 1982, p. 66). There has been much public apprehension about and opposition to siting such facilities, although Passant indicated that he was 'confident that people will accept these things' (1983, p. 11). When asked what he thought about the possibility of not having acceptable disposal routes, he answered, 'I think if we were faced with a situation where no disposal routes were available for any of these wastes one would certainly seriously rethink decommissioning policy ... ' (Passant, 1983, p. 28).

Mr Passant's remarks at Sizewell constituted one of the few references to the relationship between decommissioning and waste disposal. The relationship has not gone unnoticed by others, however. Resnikoff, a witness at Sizewell, considers that waste from retired reactors is 'a sleeping giant' (Chapter 8 this volume). Citing US Nuclear Regulatory Commission estimates (US NRC, 1983), he said that four years of decommissioning would produce a volume of waste comparable to that produced during 30 years operation of the reactor.

Thus, no one knows where the waste is going, when and how much will be moved, or by what routes it will be transported. Without this knowledge it is impossible to assess the commitment of land or to plan securely for its use.

Planned land use

A nuclear power plant, whether operational, in safe storage, or removed, will have implications for land use planning. These implications are associated with mandated safety zones, emergency evacuation routes, normal routes for the transport of materials and people, and so forth. Even if the plant is no longer operating it will have a substantial impact on how the land is likely to be used or how those who live nearby or might move in wish it to be used. The land use planner will face frustrating uncertainties over not only the type and timing of decommissioning, but also whether additional plants should be located at the same site, and whether the site of a dismantled nuclear station can ever be made safe for unrestricted use. Public risk perception, and the faith in, and requirements of, regulatory authorities will influence these matters.

At Indian Point, utility spokesmen indicated that their nuclear site might be made into a park (Lee, 1983; Levy, 1983). The San Onofre site, though Federal, is likely to pass to the State of California and it will connect the two recreational beaches now straddling the power plant. The Humboldt Bay site is likely to continue with its fossil units, and then to revert to designated wetlands (Ray, 1983). Other sites, such as those in urban environments, are likely to be made available for houses. Unfortunately, planning for any of these possibilities is very difficult because decisions have not been made, and are not near, on the type or timing of decommissioning, unrestricted dose limits, or even ultimate decommissioning procedures and disposal of decommissioning wastes.

Many witnesses at the Sizewell Inquiry noted the unusual land use conditions which surround the Sizewell site (i.e. Heritage coastline, Area of Outstanding Beauty, nearby bird sanctuaries). There is also a predominance of agricultural and recreational land use as well as a relatively low population density. In their joint proof of evidence on local Suffolk County environmental issues, Suffolk County Council and Suffolk Coastal District Council (1982) enumerated the implications of building and operating a new Sizewell power plant. Their vague statements about decommissioning in relation to the site reflected the level of uncertainty within the CEGB. 'At the time of the initial response to the Sizewell 'B' proposals the Councils had no information regarding the long term implications of decommissioning for the future use and appearance of the site, and consequently they sought a condition requiring the removal of all buildings and structures, after the shutdown of the station' (p. 36). The Councils stressed the need 'to ensure the site would be cleared and landscaped in order, ultimately, to revert as far as possible to its appearance prior to the construction of the 'A' station' (p. 36). This reversion should also be as soon as possible. As noted above, the CEGB's intention is to continue to re-use the site and/or to allow Sizewell 'B' to undergo radioactive decay until approximately the year 2040. However, the CEGB made no commitment to this programme; nor did it make any commitment to a third Sizewell reactor, although this possibility was discussed on many occasions.

Without a firm commitment on the future use of the site, the timing, or the mode of decommissioning, it is fruitless to engage in future land use planning. This is a particularly vexing problem when it comes to socio-economic conditions and their associated land use implications (e.g. road construction and widening, construction of housing and service facilities). Part of the socio-economic problem is that decommissioning (depending on the type) may continue with construction (depending on the sequencing) to produce a continuous programme of disruptive activity until the year 2200.

Single/multiple sites

The number of units and their configuration on a site will have an impact on the choice of decommissioning methods and the commitment of the land. In the case of San Onofre 1, Peach Bottom 1, Humboldt Bay, and Indian Point 1, current plans do not favour dismantling, supposedly because of the disruptive effects on adjacent units (Kohler, 1980). At Sizewell, a MAGNOX reactor has been operating since 1966, and a 'C' unit has been discussed. As long as the utility companies have fences, security, and trained personnel in place for the operating units, monitoring an adjacent nuclear plant in safe storage constitutes a relatively small incremental cost. Single-unit sites are more likely to be dismantled earlier, thus releasing the land for other uses. However, once a decision is made to build more than one unit at the same site, particularly if it is a phased rather than a concurrent construction programme, the land commitment is extended and intensified. Furthermore, land use planning becomes more complicated, and the potential use of the land for purposes other than power generation decreases.

Conclusions

Attention given to nuclear power by environmentalists has for years focused on the hazards and other impacts of the middle, or operational, phase of the nuclear fuel cycle. The neglect of back-end considerations began to appear when waste disposal needs grew more pressing. Now that first-generation nuclear power stations are nearing the end of their operational lives, decommissioning is emerging as another back-end issue not fully anticipated or assessed. The matter of decommissioning is beginning to pose difficulties largely because the full life cycle of power plants was not considered when initial nuclear planning was taking place; it was simply accepted on faith that any future problems could be solved. This approach has failed because it is essentially asocial. As we have seen in the opposition to waste disposal schemes, the public is not willing simply to accept the analysis, conclusions or decisions of the nuclear industry. Decommissioning planning is now taking shape in a similar fashion.

This chapter has shown that siting, decommissioning, and land use are related issues. It has also demonstrated that none of the relationships between these issues contributed substantively to siting decisions, although their influence could be significant. Indeed, it would be sensible to make decommissioning a mandatory siting consideration, statutorily enforced as an element in choosing future nuclear sites. It should also be a factor in any normal environmental impact evaluation.

The issue of siting has been used here as an illustration of the complex ramifications of the overall topic of decommissioning. These complexities are not limited just to power plants, however; they extend to all facilities throughout the nuclear fuel cycle. A similar examination of other phases in this cycle would reveal more completely the long-lived commitments, both technical and social, stemming from nuclear decommissioning decisions.

REFERENCES

BAKER, E. L., WEST, S. G., MOSS, D. J. and WEYANT, J. M. (1980) 'Impact of offshore nuclear power plants: forecasting visits to nearby beaches', Environment and Behaviour, 12:367-407.

BLAKE, J. (1985) Transcript of Proceedings, Day 314, 18th January, Sizewell 'B' Public Inquiry.

CLARK, J. H. (1983) Long Range Plan for Decontaminating and Decommissioning Excess Contaminated Facilities at the Idaho National Engineering Laboratory, PW-W-79-005, Rev. 4, Waste Management Programs Division, EG&G Idaho, Inc.

DIRCKS, W. J. (1982) Pressurized Thermal Shock (PTS) Policy Issue (Notation Vote), SECY-82-465, plus enclosure A: NRC staff evaluation of pressurized thermal shock.

DOBSON, J. E. (1984) 'Introduction to Section II - spatial assessments and impact mitigation' in M. J. PASQUALETTI and K. D. PIJAWKA, (eds.) Nuclear Power: Assessing and Managing Hazardous Technology, Boulder, Colorado: Westview Press, 93-101.

FELDMAN, C. (1983) 'The impact on low-level radioactive waste from decommissioning commercial light water power reactors', Memorandum for

State officials, US Nuclear Regulatory Commission.
FELDMAN, C., CALKINS, G. D., CARDILE, F., MATTSEN, C., MOORE, E., HALE, V., MURPHY, E. E. and KIEL, G. (1981), Draft Generic Environmental Impact Statement on Decommissioning of Nuclear Facilities, NUREG-0586, Washington DC: US Nuclear Regulatory Commission.
GAMMON, K. M. (1982), CEGB Proof of Evidence: On Site Selection and Site Specific Aspects, Sizewell 'B' Public Inquiry, CEGB.
GREGORY, A. R. (1982a) CEGB Proof of Evidence: on Decommissioning, Sizewell 'B' Public Inquiry, CEGB, November 1982.
GREGORY, A. R. (1982b), Transcript of Proceedings, Sizewell 'B' Public Inquiry, Day 202.
GREGORY, A. R. (1984), Transcript of Proceedings, Sizewell 'B' Public Inquiry, Day 273, 13th October.
HANSIS, R. (1980) 'Emergency planning and nuclear power plant siting in northwestern Indiana', paper presented at the 76th annual meeting of the Association of American Geographers, Louisville, Kentucky.
HINE, R. E. (1980), Decontamination and Decommissioning of the Organic Moderated Reactor Experiment Facility (OMRE), EGG-2059, EG&G Idaho, Inc.
KOHLER, E. J. (1980) 'Decommissioning Peach Bottom No. 1', presented at Pennsylvania Electric Association Power Generation Committee Winter Meeting, February 7-8.
LEE, M. (1983), Consolidated Edison Co., telephone interview, July.
LEVY, L. (1983), Northeast Utilities Co., telephone interview, July.
MILLER, R. L. (1983), Battelle Northwest Laboratories: personal communication.
NADER, R. (1981), Speech delivered at the Second Annual Humboldt Decommissioning Conference, 11 July.
OPENSHAW, S. (1980), 'A geographical appraisal of nuclear reactor sites', Area, 12, 287-290.
OPENSHAW, S. (1982a), 'The geography of reactor siting policies in the U.K.', Transactions, Institute of British Geographers N.S.7:150-162.
OPENSHAW, S. (1982b), 'The siting of nuclear power stations and public safety in the UK', Regional Studies, 16 (3), 183-198.
OPENSHAW, S. (1986) Nuclear Power: Siting and

Safety, London: Routledge and Kegan Paul.
OPENSHAW, S. and TAYLOR, P. (1981), 'U.K. Reactor Siting Policy', memorandum by the Political Ecology Research Group, in first report of the House of Commons Select Committee on Energy 1980-1: minutes of evidence. 3:990-1018. HMSO: London.
O'RIORDAN, T. (1984) 'Sizewell 'B' Inquiry and a national energy strategy', The Geographical Journal, 150(2), 171-182, July.
PASQUALETTI, M. J. (1985) 'Nuclear energy', in CALZONETTI, F. and SOLOMON, B., (eds.) Geographical Dimensions of Energy, Dorchect: D. Reidel Publishing Company, pp. 25-38.
PASSANT, F. H. (1982), CEGB Proof of Evidence: On Radioactive Waste Management, CEGB Proof of Evidence, 20th November.
PASSANT, F. H. (1983), Transcript of Proceedings, Sizewell 'B' Public Inquiry, Day 33, 1 March.
RASMUSSEN, T. L. (1977) Decontamination and Decommissioning (D&D) Plan for the Organic Moderated Reactor Experiment (OMRE), Report number WMP-77-17, Revision 1, EG&G Idaho, Inc.
RAY, D. (1983), Planner, California Coastal Commission, Eureka, personal interview, July.
RIBA (Royal Institute of British Architects) (n.d.), Submission to Public Inquiry into Sizewell 'B' Nuclear Power Station.
RICHETTO, J. (1984) 'Locating nuclear electric energy facilities in the United States: a note on structural relationships and the environment', in PASQUALETTI, M. J. and PIJAWKA, K. D. (eds.), Nuclear Power: Assessing and Managing Hazardous Technology, Boulder, Colorado: Westview Press, pp. 103-120.
RUBIN, H. (1981) 'Radioactive white elephant on the north coast', Cry California, 16, 23-28.
SAVAGE, J. A. (1979), Humboldt's First Decommissioning Conference, Sponsored by the Redwood Alliance, 3 November, transcript, Eureka, California.
SOLOMON, B. D. (1984) 'Decommissioning nuclear plants', in PASQUALETTI, M. J. and PIJAWKA, K. D. (eds.), Nuclear Power: Assessing and Managing Hazardous Technology, Boulder, Colorado: Westview Press, 387-411.
SOLOMON, B. D. and CAMERON, D. M. (1984) 'The impact of nuclear power plant dismantlement on radioactive waste disposal', Man, Environment, Space and Time (Spring 1984), 4(1): 39-60.
SOLOMON, B. D. and HAYNES, K. E. (1982) 'Multi-

dimensional programming methods for energy facility siting: alternative approaches', Proceedings of the National Conference on Energy Resource Management, Washington: NASA Conference Publication 2261, 393-409.

SOLOMON, B. D., HAYNES, K. E. and KRMENEC, A. J. (1980) 'A multi-objective power plant location model with hierarchical screening: nuclear power in northern Indiana', Review of Regional Studies, 10(3): 9-22.

SUFFOLK COUNTY COUNCIL and SUFFOLK COASTAL DISTRICT COUNCIL (1982), Proof of Evidence on Local Environmental Issues, Sizewell 'B' Public Inquiry.

TUTTLE, D. C. (1982) 'The history of erosion at King Salmon from 1854 to 1982', prepared for presentation at the Humboldt Bay Symposium, 26 March.

US COMPTROLLER GENERAL (1980), Existing Nuclear Sites Can be Used for New Power Plants and Nuclear Waste Storage, report to the Subcommittee on Energy and Power, House Committee on Interstate and Foreign Commerce of the United States, EMD-80-67, Washington DC: U.S. General Accounting Office.

US NRC (Nuclear Regulatory Commission) (1983) S.N. Solomon, Memo for State Officials, Washington DC, 10 March.

US NRC (1985) Decommissioning criteria for nuclear facilities - proposed rule, Federal Register, 50 (28): 5600-5625, Washington DC.

ZICHELLA, C. (1983) 'More delays on Humboldt nuke', Nuclear Free Times, Winter: 1,4. Published by the Redwood Alliance, Eureka, California.

ACKNOWLEDGEMENT

The US portion of the research for this work was supported by a grant from the University Research Fund of Arizona State University, Tempe, Arizona. K. David Pijawka and Paula McClain were members of the original research team, and they both made helpful contributions in the early stages of the work reported in this article. Responsibility for conclusions and recommendations, however, remains my own. I also thank Prof. Ronald Cooke, Chairman of the Geography Department University College, for his assistance in arranging for typing and cartographic work while I was on assignment in London.

Part 3:

RISK AND IMPACT: THE SOCIAL DIMENSION

Chapter Ten

THE SELLAFIELD CONTROVERSY: THE STATE OF LOCAL ATTITUDES

Sally Macgill and Siân Phipps

Sellafield: a key focus for nuclear debate in the UK

November 1983 saw a particularly controversial episode in the history of civilian nuclear power development in the UK, with a striking coincidence of events centred on British Nuclear Fuels' reprocessing plant at Sellafield in Cumbria - an installation better known to people outside the area as Windscale. At the beginning of the month, a controversial television documentary 'Windscale, the Nuclear Laundry' was broadcast, heralded by an unusual amount of advance publicity. The programme drew attention to the fact that Sellafield discharges greater quantities of radioactivity than any other installation in the UK, that land near the plant has up to one hundred times the natural amount of background radiation, that radiation is the only known cause of leukaemia in children within the limits of present knowledge, and that the incidence of leukaemia in children in the vicinity of Windscale was ten times the national average. Viewers were left to draw their own conclusions. Within two weeks of the programme, a serious operational error at the plant (referred to below as BNFL - British Nuclear Fuels Ltd) led to the discharge into the sea of a tank of radioactive effluent which should have been held for further reprocessing; part of it was subsequently washed back by the tides onto local beaches. At a later court hearing BNFL was found guilty on four charges related to this incident. The discharge in November 1983 coincided with a well-publicised attempt by the environmental organisation Greenpeace - as part of a longer running campaign against the build-up of man-made radioactivity in the environment - to block the outlet of the pipeline through which radioactive effluent is routinely discharged from Sellafield into the sea. It was Greenpeace who first made public the discharge. The unlikely coincidence of the most serious operational error in the history of operations at BNFL with the Greenpeace protest has been widely noted: some have surmised that other operational errors may have occurred but simply gone undetected; others, though, point to it merely as an unfortunate incident in the company's otherwise good record.

The speed of Government response to the events of November 1983 bears witness to the sensitivity of the issues forced into wide public view. The Government called for increased monitoring of radiation concentrations in the area; independent studies into the contamination of beaches around the plant (Department of the Environment, 1984; Health and Safety Executive, 1983); a statement from the Department of the Environment (apparently to the dismay of local councils) advising the public not to make 'unnecessary use' of a twenty mile stretch of local beach owing to possible radiation contamination; and a high-level inquiry, under the eminent medical scientist Sir Douglas Black, into the possible increased incidence of cancer in the vicinity of the Sellafield site. This inquiry would stand as the foremost expert testimony on the highly controversial issue of whether children in the vicinity of the plant are environmental victims of nuclear power. The epidemiological evidence for this inquiry is examined elsewhere in this book by Openshaw and Craft (Chapter 11). The purpose of Sir Douglas Black's inquiry has been widely interpreted as being partly to 'get at the facts' and partly to assuage public opinion (Black, 1984).

To many, the events of November 1983, and subsequent responses, are uninteresting and irrelevant. To others they represent small but ultimately insignificant aberrations in the otherwise reliable operations of BNFL at Sellafield. But to some they represent ugly evidence of a dangerous industry whose operations cannot be adequately controlled and should be drastically, if not completely, curtailed - merely the latest in a number of blatantly unsatisfactory aspects of development and operations at the Windscale site. These include a fire in 1957 (British Nuclear Fuels, 1983), military connections of uncertain extent and significance (Bunyard, 1981), a controversial planning inquiry in 1977, and routine discharges of radioactivity that are several orders of magnitude higher than those of comparable reprocessing plants elsewhere in the world.

Why study local attitudes? Aims and justifications

Research into public attitudes may be justified in two main ways. Firstly, it can be of intellectual value. Looking at how people think, feel and express themselves on particular issues contributes to our knowledge and understanding of mankind. Complementing the intellectual argument, research into public attitudes is also useful to the decision makers in both government and industry. To know about local feelings on particular environmental issues - the feelings of people on whose behalf they are supposedly acting and in whose neighbourhood they are operating - might help them towards better public relations. Or at the very least, it might demonstrate to institutions of authority the complexion of public thinking on ostensibly technical issues. As Mitchell (1979) points out, resource managers are not always well attuned to the perceptions of the general public; their views on how 'the public' interprets and reacts to the environment may greatly differ from the public's own perceptions of things. Similarly:

> Chemists, climatologists, physicists and others who might tread very gently beyond the available data in their own areas of specialisation at times feel no restraint in opining that "this is what lay people think about the risk of nuclear power", that "this is the sort of information the public needs to put risks into proper perspective" or that "this is how the public will react to stricter safety measures" (Fischoff et al., 1981).

In the light of these considerations the present research aims to elicit and interpret attitudes of local people on key issues at the heart of the Sellafield controversy. Before it was undertaken there had been no attempt to complement the radiophysical, radiobiological, epidemiological and medical inquiries that had been completed or were under way into radiation concentrations and consequent potential health risks around the Sellafield site, by examining the feelings and perceptions of local people. It is these people who are living in the area disputedly 'at risk', and who, as communities, have most closely experienced the impact of the BNFL plant and indeed the associated publicity. In other words, our aim was to present the social-psychological setting that has not been given a place in other studies that have been carried out or commissioned.

The Sellafield case then, provides both a context for adding to existing empirical knowledge about the perception of risk, and for addressing its relevance in an important policy setting. Our empirical survey of attitudes involved not only determining the spread of opinion on particular issues, but also the identification of patterns which might exist in the responses and how these might be accounted for. It is, after all, the recognition of patterns and not simply the accumulation of isolated, unrelated findings that will contribute to the development of knowledge about perceptions. The survey should also put media interpretations of local opinions, interpretations which by nature are individualistic and often extreme, into better perspective. The image portrayed in much media coverage had been one of local unrest towards BNFL and a great deal of local anxiety about health risks (most notably in the crucial YTV documentary). Against this and the Black Inquiry's evident aim of seeking to reassure local people, we deemed it appropriate to provide some indication of the real extent of local agitation and the local need for reassurance.

At the same time, however, there were related aspects that should not be counted among the aims of the research. We were not seeking to develop fundamental theoretical insights into risk perception - our aims were less ambitious than that and our tools of inquiry lacking in sufficient power and sophistication. We believed there was worthwhile knowledge to be gained from conducting a more limited type of social survey in the area. Neither was it intended to moralise on the 'rationality' or 'accuracy' of local opinions: there are no 'right' or 'wrong' views when it comes to questions of risk perception - each person's opinion should be seen as legitimate since it embodies very real individual needs and values. The intention was merely to make the sentiments of the (previously unconsulted) 'silent majority' known and to interpret them.

The <u>balance</u> of opinion that might be found on certain related issues was impossible to predict. On the one hand, Sellafield has a unique position in the nuclear fuel cycle (mainly reprocessing of spent fuel - a relatively labour intensive operation) and has a vital importance to the local social, economic and political fabric. But on the other hand, it has also a unique position as the subject of considerable publicity surrounding the November 1983 discharge and the allegation of an excess of childhood cancers in the plant's vicinity.

Whereas expectations of the survey findings had to remain open, the design of the survey questions had been very much conditioned by conceptual influences from other sources. These influences - from theoretical and empirical literature, media representations and background knowledge of the area - are outlined in the following sections.

Research into public opinion towards nuclear power

The state of public opinion on nuclear power has been increasingly researched in recent years. This research has been partly general - 'nationwide' counts of broad 'pro' and 'anti' views about the development of nuclear energy - and partly more specific - on views of local people about existing or proposed nuclear facilities. Considerable research has been devoted simply to generating raw opinion polls - crude attitudinal barometers - with no pretence of making a theoretical contribution to the field of risk perception research. Others have tried explicitly to develop theory on risk perception. A lot of research for the raw opinion polls has been sponsored by agencies with an active interest in the outcome of particular decisions as a means of providing useful information on how they are perceived by others. Some of this work is reviewed below by way of general context for the new empirical findings presented later in this chapter.

The literature suggests factors that might account for different patterns of perception, though, as we have remarked, it gives little away as far as 'predicting' opinion around Sellafield is concerned. We are addressing a particular public here - the people of West Cumbria - and asking our own particular questions: it has been pointed out elsewhere that even apparently minor changes in wording can be important in yielding significantly different responses.

According to van de Pligt *et al.* (1984a),

before the mid 1970s, survey data showed consistently high levels of support for nuclear power. However, since the mid 1970s, support has gradually waned. Referenda in Switzerland and Austria were all decided with very small majorities in favour of nuclear power, while in the Netherlands a majority have been anti-nuclear since the mid 1970s. UK opinion polls since the mid 1970s show a slow but steady increase in public opposition to nuclear energy. A national poll in October 1981 showed 33% in favour of expanding the number of nuclear power stations in the UK, 53% were opposed (van de Pligt _et al_., 1984a).

Reduced support for nuclear power might be superficially described as an indication of increasing 'distance' between the desires of planners in the nuclear field and the wishes of the public (Brown, 1985). It might more satisfactorily be interpreted as bound up with a crisis of legitimation between nuclear planners and the public: part of a wider crisis since the War (Kemp, 1980).

As nuclear power has emerged as an important issue for public debate so there has been an increase in media coverage. Nealey _et al_. (1983) suggest that in America media coverage emphasises the disadvantages rather than the benefits of nuclear power, and in turn is more likely to influence people in an anti-nuclear direction. Individual adverse events obviously contribute to such a trend, notably the Three Mile Island accident in the USA which was followed by a decrease in nuclear support. Nealey _et al_. (1983) suggest, though, that despite the perception of attendant risks, people do not wish to forego the nuclear option, nor to lose out on jobs associated with investment. See also van de Pligt _et al_. (1984b).

Psychologists have sought to develop different theories of why opinions towards nuclear risks should vary (between different people and at different times). The following summary of key factors draws on a recent contribution by Lee to a Study Group report from the Royal Society (Lee, 1983).

1. Knowledge. People have different levels of knowledge about what the physical risks actually are, or how they might be estimated - perceptions may be too conservative or too alarmist as a result. A seminal report by the Council for Science and Society (1977) suggests that the most

acceptable risk is perhaps the one about which people are most ignorant, and Douglas and Wildavsky (1982) also suggest that the very dangers people most seek to avoid are those that may harm them the least.

2. Compensation or benefit. People stand to benefit in different ways from activities that might be deemed to be sources of risk but bring jobs, indirect income, useful consumer products or security to a neighbourhood. Risk, therefore, cannot be separated from the benefits or any other characteristics of the industry or technology with which it is associated, and an individual's perception of risk can only be properly understood in a wider context.

3. Culpability. This heading is used to embrace two related characteristics, namely the degree of willingness of a person to accept exposure to a possible risk and the degree of control a person has over a perceived source of risk. (For example, people voluntarily indulging in 'risky' sports may expose themselves to the same level of physical risk as people 'involuntarily' living close to a hazardous chemical store but their evaluation of risk may be quite different.) How people see such factors can be fundamental both in acting as a catalyst for deeper-seated risk concern and in seeking apportionment of blame in the event of accident.
   Otway and Thomas (1982) give a telling account of the culpability characteristic, arguing that mistrust in the institutions responsible for controlling a risk source should not be regarded as a symptom of, but rather as a catalyst for, related risk concern.

4. Familiarity. The Royal Society Study Group's Risk Report (Lee, 1983) notes that it has been amply demonstrated at the neurophysical level that living organisms react less and less to all stimuli with repeated exposure (a familiarity effect). It is an adaptation effect thought to depend on the fact that repeated exposure contains no intrinsically new information.

They suggest that this explains why anxiety towards nuclear plants may increase with distance from such plants, although alongside this explanation another factor is introduced - that rumours (products of low information and a cause of inaccuracy) may be more primitive also at further distances.

5. Other risk. People's differing tendencies to compare a given risk with other risks (and in each case with the wider characteristics of the risk-generating activity or technology) may also correlate with different perceptions - people having different points of reference in terms of what they find acceptable.

6. Mental capability. Individual abilities to visualise models of cause and effect, to conceptualise one or many aspects of a risk context and to differentiate and synthesise these aspects, clearly differ. How it all impinges on their wider values, feelings and goals (self-interests) in life will likewise vary enormously.

7. Social contexts. Different people subscribe to different social norms (working-class norms, middle-class norms) and face different social pressures. The normative component is well established as an important determinant of environmental cognition, and although forces of social obligation are recognised to be crucial, O'Riordan (1981) suggests that they can be most difficult to unravel (e.g. the influence of friends, peer groups and other significant social networks). Individual opinions or 'expert' views may, therefore, be deliberately distorted in order to fit the norms of the respected social group.

Such factors, then, begin to account for the very different attitudes that various individuals may have towards a given source of risk, and the second, third and fourth were to be explicitly taken account of in our survey design. The same factors have also been considered elsewhere in relation to the dichotomy that exists between 'objective' and 'perceived' risk: in other words, to account for the dissonance often observed between people's concern or worry about a particular risk and the best technical estimates of potential or actual physical harm. This dissonance has been 'explained' in terms of such factors as the seven listed above, and others - such as whether the effects are immediate or delayed (e.g. Slovic *et al.*, 1980). There are different views as to whether a 'rational' perspective is one based specifically on best technical estimates of physical harm, or whether rationality can itself only be defined with respect to a particular world view, within which psychological dimensions have their place. The most penetrating appraisal of this issue has, to date, not come from psychological constructs as such, but from adopting a more overtly cultural stance in which cognitive, physical and social dimensions of existence are more closely bound together. Such considerations take us somewhat beyond the domain of the present chapter (but see Macgill and Berkhout, 1985), which is concerned with the particular complexion of attitudes and opinions within West Cumbria, as expressed in the wake of publicity and controversy following the striking events of November 1983.

Sellafield in its West Cumbrian setting

The Sellafield site is on the coastal plain of West Cumbria, bordered by the Irish Sea to the west and, a few miles inland to the east, the mountains of the Lake District National Park (see Figure 10.1). It is not a particularly accessible part of the country, the topography inhibiting transport links with the rest of the UK and contributing (from an outsider's viewpoint) to the similarly detached nature of local humour and outlook. Sellafield is the dominant economic unit and the primary social institution in the area. It currently provides about 10,000 jobs - a large industrial concern by most standards - and because of the intensive nature of reprocessing operations in comparison to nuclear power generation it is a far larger employer than any other nuclear installation in the UK. It is also an exceptionally important component of the industrial base of West Cumbria - the former basic industries of iron and coal being well into decline. Roughly two-thirds of the jobs at Sellafield - a dominantly male employer - are permanent, with a marked polarity between highly skilled and

Figure 10.1: LOCATION MAP: SELLAFIELD AND NEARBY TOWNS AND VILLAGES. SURVEY UNDERTAKEN IN PLACES UNDERLINED

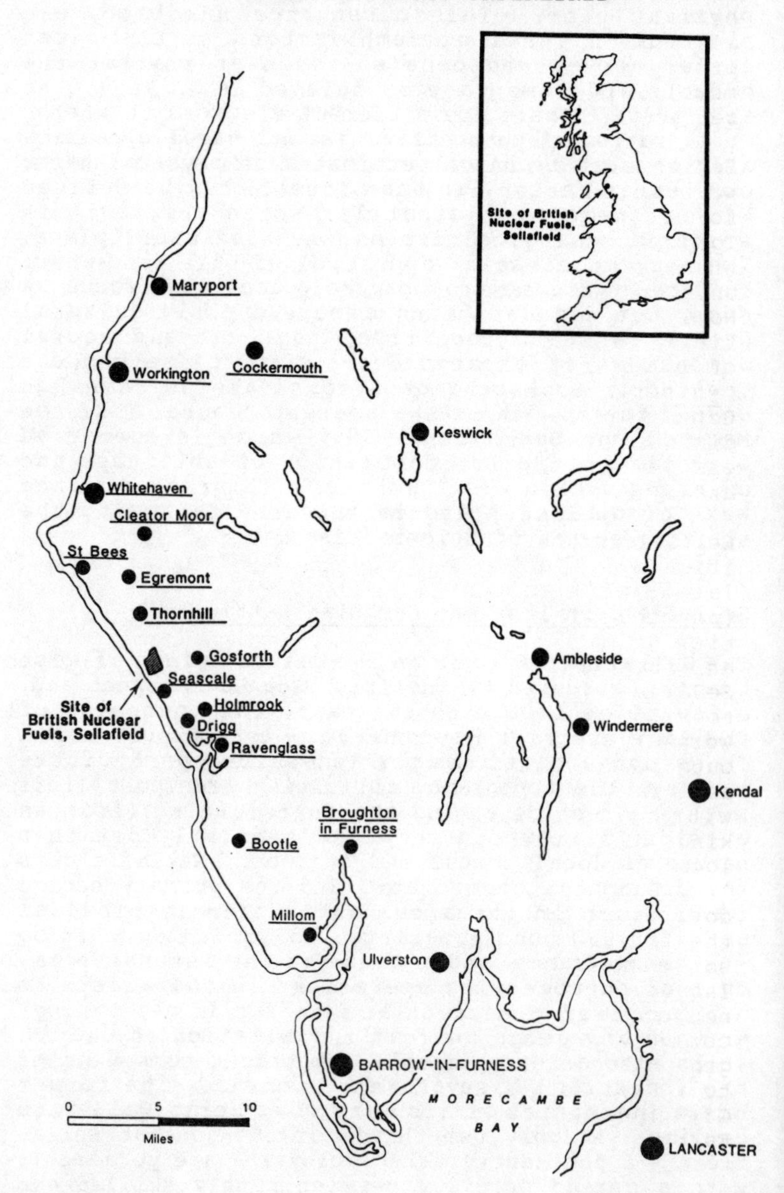

unskilled workers: but about 4,000 are temporary construction posts. Considerable expansion and redevelopment is now taking place on the site - for example, the thermal oxide reprocessing plant. In addition to the direct employment which Sellafield provides, there are also considerable indirect multiplier effects - many hundreds of jobs in local service industries for example.

Many local villages have high proportions of Sellafield employees, most obviously Seascale which was built largely in the 1950s to house the workforce for the original Windscale atomic piles. Towns further away also house Sellafield workers - Whitehaven, Cleator Moor, Maryport, Millom and so on, but these also have their own local industries. Notwithstanding the large number of jobs at Sellafield, there is concern that the greater proportion of Sellafield jobs should be given to local people (Whitehaven News, 1985); the unemployment rate of Copeland District (Sellafield's local authority district) is over 10% (1981 Census), and many workers at Sellafield come from outside the immediate area.

It is not only the economic importance of Sellafield that should be acknowledged: the identification of a large proportion of the local population with BNFL will undoubtedly have a bearing (psychologically and culturally) on their perception of the issues covered in the questionnaire - an effect which, in view of the distinctive role of Sellafield in its locality, would not be expected to exist elsewhere in the UK. For most people who work at BNFL, perceptions of disputed health risks outside the plant can hardly be divorced from their association with work practices, day-to-day affairs within the plant, and the filtering of different kinds of knowledge - by word of mouth - through the local community.

Such publicly expressed concern and debate as does exist occurs mainly through the local press, the Parish Councils and the Sellafield local liaison committee, established after the Windscale Inquiry of 1977 in order to provide a better channel of communication between BNFL and the local community. Further from the plant environmental groups are active both in Barrow (Cumbrians Opposed to a Radioactive Environment - CORE - affiliated to Greenpeace), and in Maryport (Friends of the Earth).

Survey design

A questionnaire was designed to ask local people about their attitude towards BNFL, about the extent of their concern (or lack of it) for radiation levels and health risks that might be associated, and about related issues that might have a bearing on these views (see pages 222-224), or changes in these views.

The survey was undertaken in April 1984, three months before the Black Report was published and when the legacy of the November 1983 events was much in evidence in the area (e.g. in the local press). The survey area included all large villages within ten miles (by road) of Sellafield and larger towns further afield up to twenty miles. The places covered by the survey and the population sampled are indicated in Figure 10.1. The method of sampling was conditioned by available resources - twelve interviewers (final year undergraduates, specialising in energy and environmental subjects), and two days in which to complete the work. The intention was to sample in proportion to settlement size. It was simply not practicable to select randomly from the electoral register; instead, as a second best, interviewers were assigned a particular area, per day or per half day, and instructed to approach a specified number of homes in that area selected to reflect (if applicable) differences in housing stock. If all approaches had been successful and all the interviews had taken the same time, samples proportional to each settlement size would have emerged. In all, 395 residents were approached, 263 (67%) of whom completed a questionnaire. Reasons for refusal to complete the survey, which, in a sense, can be read as part of the general survey findings are given in Table 10.1. Some interviews were lengthy (up to two hours) which, while interesting and educational for the interviewer, meant that fewer than the desirable number of households could be covered in the time available. In this chapter, therefore, we do not regard spatial or social factors as statistically significant: the numbers are too small. Rather, we present aggregate views of a sample of West Cumbrian people. The age distribution of the respondents to the survey is given in Table 10.2; 60% of the respondents were female.

Table 10.1

REASONS FOR NON-PARTICIPATION IN THE SURVEY

Reason given	Number of times
Too busy	23
Not interested in the survey	16
BNFL employees who thought it better not to give their view	13
Considered that it didn't concern them	11
"There's nothing wrong at Sellafield: there has been a fuss over nothing"	10
Wives of BNFL employees who thought that they ought not to answer these sorts of questions	8
Did not want to talk about it	6
Inferred that they were "sick of it"	6
No opinion on these matters	6
Considered themselves insufficiently informed to express a view	6
Nothing to say	4
Thought their husbands should fill in the survey	2
Thought it too controversial	1
Too frightened	1
Did not want to talk to an environmental group	1
No reason given	18

Figure 10.2 ATTITUDE TO BNFL AND EFFECT ON HOUSEHOLD LIVELIHOOD

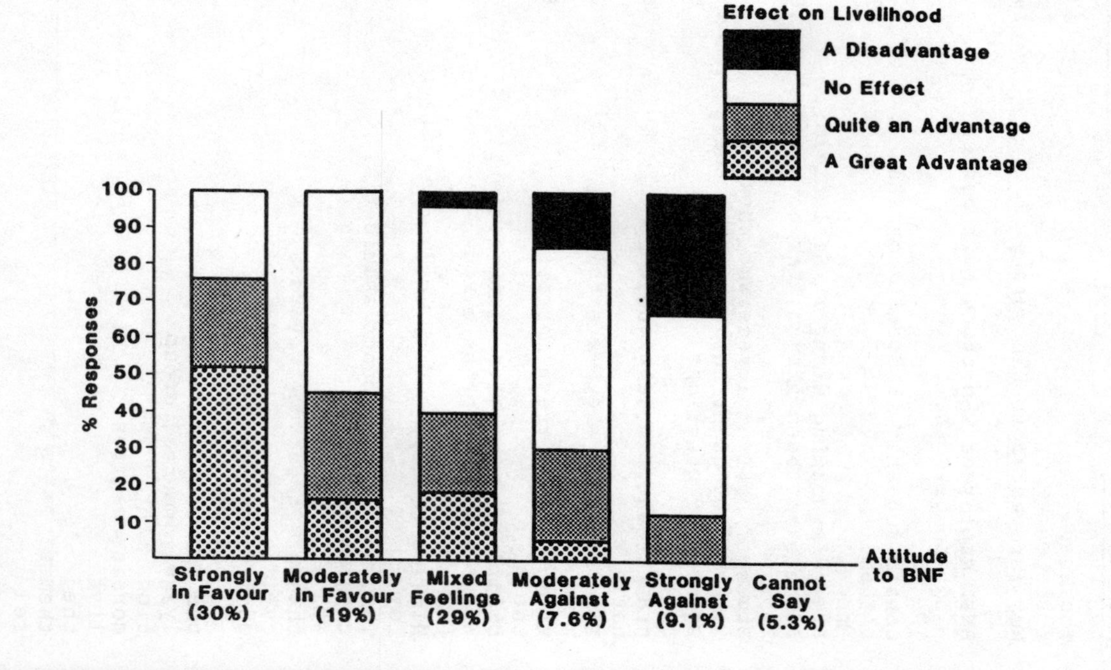

Table 10.2

DISTRIBUTION OF RESPONDENTS BY AGE

Age (years)	16-25	26-35	36-45	46-55	56-65	65+
% respondents	18	22	22	13	11	14

Attitudes towards BNFL

The first question asked people about their general attitude towards BNFL as a local company, recording their response as either 'strongly in favour', 'moderately in favour', 'mixed feelings', 'moderately against', 'strongly against', or 'cannot say'. Approximately half of the respondents were favourable towards BNFL either moderately (9%) or strongly (30%); many had mixed feelings (29%); while the minority were either moderately or strongly against (8% and 9% respectively). This significant local support for BNFL stands in contrast to the impressions of local feelings given by national media. Indeed, the wider national public may be surprised to learn of such pro-BNFL sentiments in West Cumbria. The stability of local views towards Sellafield was also significant, though there may be reservations about the reliability of responses to retrospective questioning. When asked about their views twelve months before (i.e. before the YTV documentary, the November 1983 discharge and all the adverse publicity that followed) people showed few real changes in outlook. Then, slightly fewer had been strongly against the company (2%) or had mixed feelings (about 7%), while slightly more people had been in favour of the company. It appeared, therefore, that the events of the previous year had had little effect on people's general opinion of BNFL.

Such stable support must partly tie in with the heavy dependence (either directly or indirectly) of local livelihoods on the continued operations of the plant. Nearly half of the respondents considered that the plant was an advantage to their livelihood; a mere 5% considered BNFL to be to their disadvantage; while the remainder believed themselves to be unaffected by Sellafield. A close relationship between attitude towards BNFL and a person's economic dependence on the company was borne out by cross tabulating the responses to

Figure 10.3 ATTITUDE AND GENDER

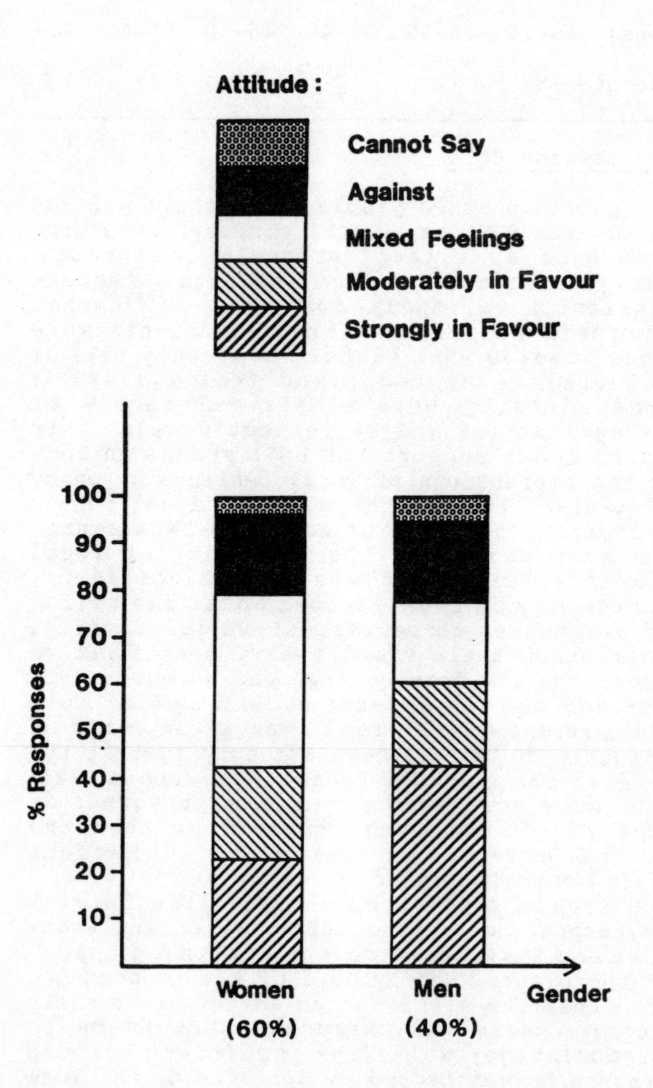

these two questions. As can be seen from Figure 10.2, most people who do not depend on BNFL have 'mixed feelings'; those who consider the company to be an advantage to them are overwhelmingly (but not unanimously) in favour of BNFL; while the very few who believe it to be a disadvantage to their livelihood are against the company.

A more striking dimension of local attitudes was the differences between men and women respondents. Approximately two in every five men were 'strongly in favour' of BNFL - only about one in every five women expressed such a view (Figure 10.3). But over a third of women interviewed had 'mixed feelings', an attitude shared by less than one in five men. In helping to explain these gender differences it is useful to reiterate the employment structure of the area. BNFL has a predominately male labour force: because it is their job and their livelihood men may (consciously or subconsciously) rationalise their financial interest in terms of a positive attitude towards the plant. Fewer women, on the other hand, find themselves in such a position - a gender pattern which also emerges in the context of risk concern.

Local concern

Local opinion is both complex and diverse on the subject of concern for possible radiation risks. We addressed two particular aspects of this subject - concern for the health of different members of the community, and concern for various possible sources of radiation concentrations.

For the first aspect, people were asked about their concern (or lack of it) for themselves, and for local children. Some definite priorities of concern emerged - there was less concern for personal risk than for the risk to local children. Here the predominant view was of no worry at all (55%). There was considerably greater concern for local children; one in five people said they were 'very worried' for them (see Table 10.3). Such findings put into perspective media representations of local concern - notably the YTV documentary broadcasting very emotional interviews with local householders.

More concern was expressed about possible sources of radiation contamination (Table 10.4). One-third of respondents stated that they were 'very worried' about radiation levels on the bea-

Table 10.3

HOW WORRIED ARE PEOPLE ABOUT RADIATION LEVELS IN THE AREAS AROUND BNFL's PLANT AT SELLAFIELD AS REGARDS HEALTH RISKS TO THE FOLLOWING PEOPLE?

	Not at all	Slightly	Moderately	Very	Cannot say
themselves (% responses)	55	22	12	11	1
local children (% responses)	35	26	16	19	2

ches - a higher percentage figure than for personal risk or risk to children (c.f. Table 10.3), although nearly a quarter were 'not at all worried'. About a fifth of respondents were 'very worried' about radiation levels in seafood (and two fifths were 'not at all worried' here), while most people were not concerned about the air, milk and house-dust as conceivable sources of radiation concentrations (50% were ' not at all worried' for each of these).

Again the differences between men and women were striking (see Figure 10.4). Women were more acutely concerned . Given the different roles of men and women - in this area especially in terms of economic activity - the deeper significance of gender may well be a fruitful area for more detailed research. As far as the Black Report is concerned, it appears that it was women in West Cumbria who were most in need of Sir Douglas Black's 'qualified reassurance' (para. 6.13, Black, 1984).

A definite relationship was found between people's level of concern and their economic dependence on BNFL: the more dependent a person was on the company, the less likely he or she was to be worried about radiation risks. Similarly, it was not surprising to find a correlation between people's concern for possible radiation risks and their attitude as a whole towards BNFL: a high percentage of those 'not at all worried' were strongly in favour of the company, and high proportions of those 'moderately' or 'very worried' had mixed feelings or were moderately or strongly against. It would be interesting to know more about how people consciously (or unconsciously)

Table 10.4

HOW WORRIED ARE PEOPLE ABOUT RADIATION LEVELS IN EACH OF THE FOLLOWING

% responses	Not at all worried	Slightly worried	Mod. worried	Very worried	Cannot say
on the beaches	23	29	14	32	2
in house dust	59	13	10	15	5
in the air	49	20	13	17	2
in sea food	43	19	12	21	5
in milk	55	16	9	16	5

rationalise their concern (or unconcern) in this respect - whether compensatory mechanisms exist to block out fears about risk; whether experience with working in the plant is bound up with the knowledge that there is no serious environmental risk; or whether lack of concern may have been part of people's justification in their particular choice of livelihood.

Information and participation

The Black Inquiry's Report would have constituted new information about the disputed environmental risks at Sellafield, and have provided a basis on which people might have reassessed or reaffirmed their views (and one which perhaps might be deemed particularly authoritative see also Macgill et al., 1985). Other sources providing members of the local public with information about Sellafield are listed on the left-hand side of Table 10.5. BNFL produces its own public information documents on

Figure 10.4 CONCERN AND GENDER

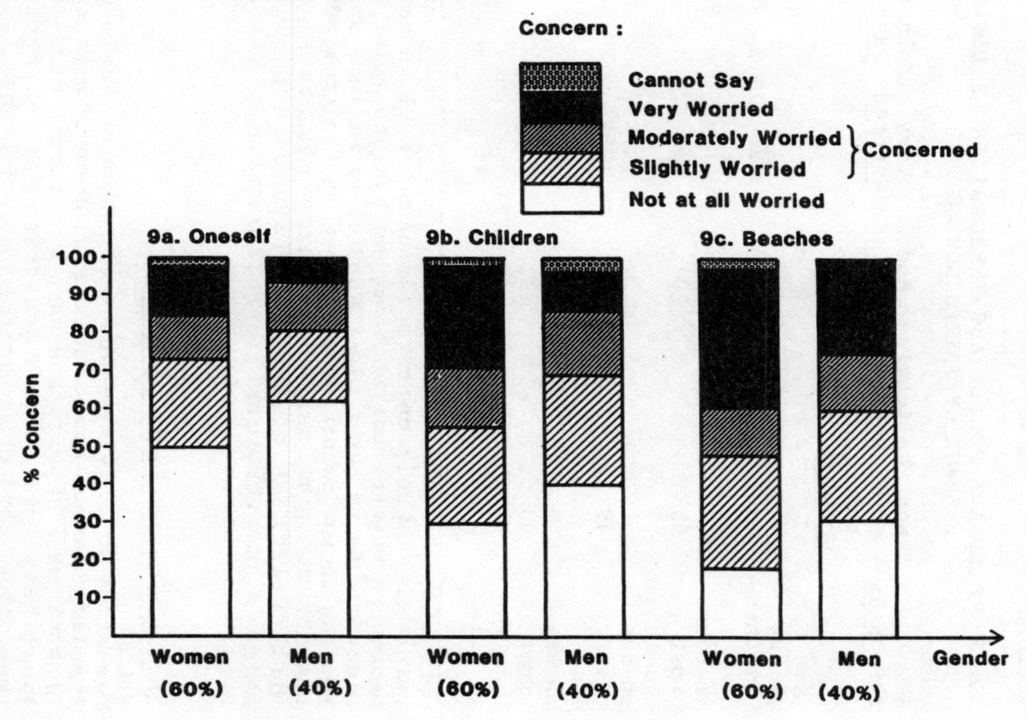

various aspects of its operations. As suggested, many local people may have further knowledge about the plant through personal contacts with BNFL employees, but many more of the public at large rely on BNFL's television image and its comments as reported by the press. The activities of the local liason committee are publicised in the local press, and its meetings are open to the public to attend. The term 'Government Bodies', as used in this Table, embraces the Department of the Environment, the Ministry of Agriculture, Fisheries and Food, the Nuclear Installations Inspectorate, Sir Douglas Black's Committee and so on - institutions which have been involved both recently and in the past with monitoring, authorisation and official decision making concerning Sellafield's operations. Their information takes the form of special government reports, but more often than not they reach local people via the media, along with editorial and other comment of the media's own design. And the other named source of information, 'Environmental Groups', includes organisations such as Greenpeace and Friends of the Earth. These publicise their activities through their own literature, but again reach the wider public via television screens and newspaper columns.

Table 10.5 illustrates the varying degrees of public confidence in these various sources of information. The picture is by no means one of unalloyed confidence in authority - BNFL may be perceived as the most reliable institution in local communities, but its record is not a totally convincing nor impressive one - more people thought it unreliable than thought it reliable; government bodies too, fare disappointingly as trustees of public faith and trust; while the local liaison committee is chiefly beyond respondents' conceptions - neither its functions nor its existence were very well appreciated.

Public faith in the institutions associated with the management and regulation of risk has been recognised as of increasing importance in understanding people's attitudes on particular issues (Wynne, 1983; Otway and Ravetz, 1984). It has been shown that the means of achieving public 'legitimacy', 'credibility' and 'acceptance' are diverse - openness and frankness are thought to engender public trust on the one hand; institutional ritual and mythology have been used to secure public faith on the other. It is important, therefore, to have information on the ways in which the public perceives the institutions of power and authority - both to gain insight into how people formulate opinions and as means of assessing the impact of policy measures.

Examining the relationship between trust in

Table 10.5 THE CONSIDERED RELIABILITY OF DIFFERENT SOURCES OF INFORMATION

% responses	completely reliable	adequately reliable	of variable reliability	unreliable	cannot say
British Nuclear Fuels	16	25	28	25	6
local liason committee	6	20	22	10	42
government bodies	10	27	26	21	16
local newspapers	7	27	38	21	7
quality press	5	22	30	18	25
popular press	2	10	29	41	16
T.V.	8	18	39	28	6
environmental groups	9	19	31	30	11

various institutions and the degree of concern people harboured yielded few surprises. Figure 10.5 illustrates the close relationship that was found between trust in BNFL as a source of information and perceived risk for local children. Most people who were 'very worried' (64%) considered BNFL to be 'unreliable'; most people who were 'not at all worried' believed BNFL to be 'completely' or 'adequately' reliable. Trust, then, does appear to be an important factor in accounting for people's attitudes towards the question of risk. It would

Figure 10.5 CONCERN FOR LOCAL CHILDREN AND DEGREE OF TRUST IN BNFL

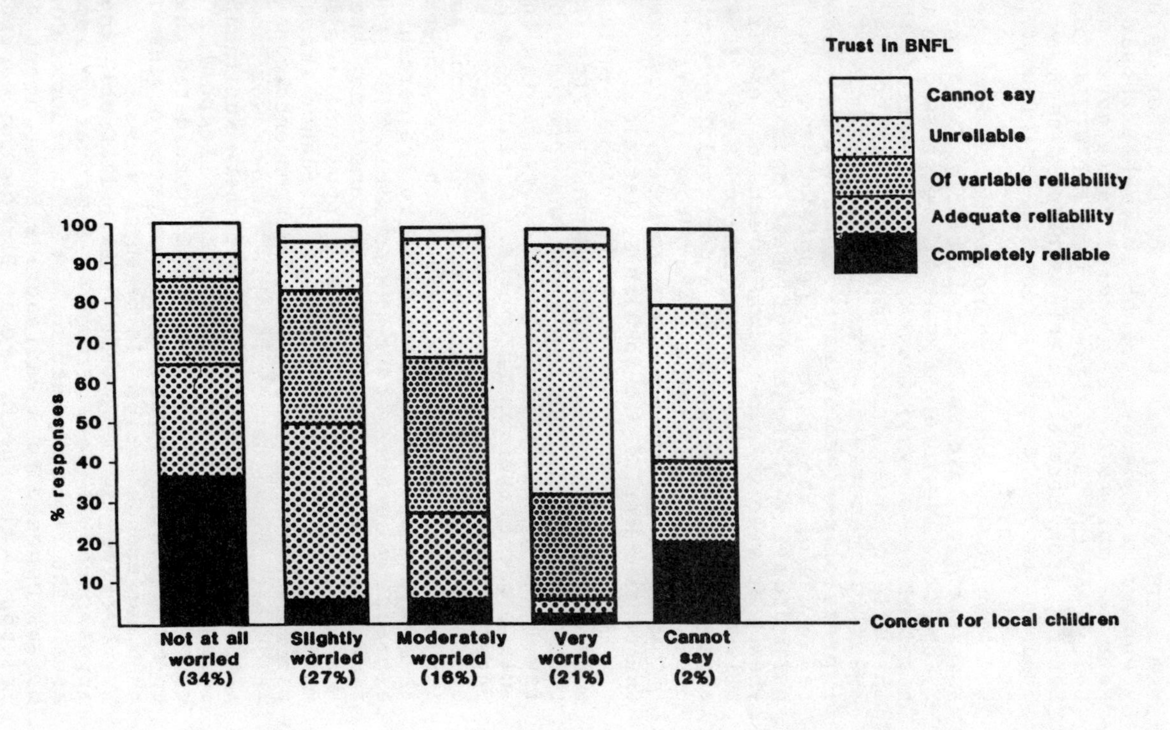

be interesting, however, to explore the mechanics of public faith in more detail - the means by which BNFL has secured local trust; whether loyalty to one's employer is a function of faith or merely economic dependence. Doubtless, BNFL should be keen to improve on its record, while government bodies and the local liaison committee would, also, benefit from more efficient means of disseminating information if they are to function legitimately in the public eye.

The competence of respondents to give their assessment of information reliability is underlain by the positive interest generally expressed in the topic of the survey; indeed, perhaps more so than in nuclear power opinion surveys carried out in areas far removed from any nuclear plant. This impression derives in part from the very evident relevance of the issues to people's lives, and in part from the abundance of additional points of view that were volunteered on aspects not specifically addressed in the survey questionnaire. It was also evident in response to a particular question on this: 90% expressing either 'some interest' or 'a lot of interest' in written articles, programmes and other materials to do with Sellafield.

While the responses in Table 10.5 may indicate some measure of criticism of authority, from further questions in the survey it was apparent that few local people translated their criticism into active participation in debate. Attending meetings, correspondence, petitions and so on, is undertaken by a minority of people. While such activities involved from 5 to 19% of the respondents, this percentage is not even close to the percentage expressing concern about the possible risks from radiation concentrations. At the same time, there was much dissatisfaction about opportunities for members of the public to exert influence through participation. Although 28% of respondents considered that the situation in this respect was 'about right', rather more (35%) thought there was insufficient opportunity for members of the public to express a view, and a further 22% considered there to be sufficient opportunity to express a view but that not enough notice was taken of views.

'Active' local opposition to Sellafield, then, appears not to come from the immediately local area. The region's leading activist groups, Cumbrians Opposed to a Radioactive Environment and Friends of the Earth, are in Barrow and Maryport respectively - some distance to the south and north.

People were also asked about their attitudes towards opportunities for environmental groups to exert influence, and a very wide range of responses was given. These were largely in keeping with responses to a related point raised elsewhere in the questionnaire, where attitudes towards the activities of Greenpeace ranged from unreserved approval to unqualified objection, with a fairly even spread - for example, over 40% thought that Greenpeace activities were necessary - a useful watchdog and catalyst for debate and action to tighten the control of radioactive build-up in the environment, although many would prefer the group to achieve their ends by somewhat different means.

Conclusions and recommendations

While we have presented some of the key findings of a local survey, attempted to provide plausible (though by no means the only) interpretations of them, and illustrated some of their relevance in a contemporary policy context, we recognise that we have touched upon only the surface of the complex of factors which generate people's attitudes and opinions. A more exhaustive survey design would be needed to investigate these views in greater depth.

If surveys such as this have a role in providing political capital to active interests in the debate, then there is something for everybody in our findings. BNFL may be assured of a bedrock of local support; its position is under no serious threat or overwhelming pressure from local communities. The Government, too, may be relieved to find that there is little sign of widespread panic or unrest in the area. On the other hand, environmental groups find common sentiments with many local people; local concern over possible radiation risks should not be ignored by authority; associated risks are, in many minds, very real and apparent. More importantly, local opinion, if it can be summarised at all in terms of the categories adopted in the questionnaire, must be represented as a spectrum of views, no individual part of which can be summarily dismissed as that of an insignificant minority.

As we see it, the need now is to improve the basis on which a more thorough assessment of local attitudes might be made. A priority in a more thorough assessment would be to adopt a different structure to the questionnaire - one which would allow respondents greater freedom of expression and, in turn, provide data from which it would be possible to gain deeper insight into the more complex issues of attitude and opinion formation. A more rigorous sampling method might also be adopted, as well as pursuing a longitudinal approach to research in order to achieve a fuller understanding - studying the Sellafield case as it develops and the effect of related events (for example, the publication of the Black Report) on public attitudes.

Acknowledgement

Siân Phipps would like to acknowledge financial support from the Economic and Social Research Council.

REFERENCES

BLACK, Sir Douglas (1984) 'Investigation of possible increased incidence of cancer in West Cumbria', Report of the Independent Advisory Group, HMSO, London.

BRITISH NUCLEAR FUELS (1983) 'Incidents at Sellafield involving abnormal release of activity into the environment 1952-83', (SDB239/2L), cited in Black, (1984).

BROWN, J. (1985) 'How the public think, feel and act towards nuclear power', Paper presented to the Institute of Energy, March. Author at the Department of Psychology, University of Surrey.

BUNYARD, P. (1981) Nuclear Britain, Harmondsworth: Penguin.

COUNCIL FOR SCIENCE AND TECHNOLOGY (1977) Of Acceptable Risk, London: Barry Rose Publishers.

DEPARTMENT OF THE ENVIRONMENT (1984) An Incident Leading to Contamination of Beaches near BNFL Windscale and Calder Works, Sellafield, November 1983, London: Radio-Chemical Inspectorate, Department of the Environment.

DOUGLAS, M. and WILDAVSKY, A. (1982) Risk and Culture, California: University of California Press.

FISCHOFF, B., SLOVIC, P. and LICHTENSTEIN, S. (1981) 'Lay and expert in judgements about risk', in O'RIORDAN, T. and TURNER, R. K.

(eds.), Progress in Resource Management and Environmental Planning, Vol. 3., Chichester: Wiley.

HEALTH AND SAFETY EXECUTIVE (1983) The Contamination of the Beach Incident at BNFL, Sellafield, November 1983, Liverpool: HM Nuclear Installations Inspectorate.

KEMP, R. (1980) 'Planning, legitimation and the development of nuclear energy: a critical theoretic analysis of the Windscale Inquiry', International Journal of Urban and Regional Research, 4, 350-361.

LEE, T. (1983) Risk Perception Risk Assessment: a Study Group Report, Chapter 5, London: Royal Society.

MACGILL, S. M. and BERKHOUT, F. G. (1985) Understanding 'Risk Perceptions'; Conceptual Foundations for Survey-Based Research, School of Geography, University of Leeds.

MACGILL, S. M., RAVETZ, J. R. and FUNTOWICZ, S. O. (1983) Scientific Reassurance as Public Policy: the Logic of the Black Report, School of Geography, University of Leeds.

MITCHELL, B. (1979) Geography and Resource Analysis, London: Longman.

NEALEY, S., MELBER, B. and RANKIN, W. (1983) Public Opinion and Nuclear Energy, Lexington, Massachusetts: Lexington Books.

O'RIORDAN, T. (1981) Environmentalism, Chapter 6, London: Pion.

OTWAY, H. and THOMAS, K. (1982) 'Reflections on risk perception and policy', Risk Analysis, 2, 147-159.

OTWAY, H. and RAVETZ, J. R. (1984) 'On the reputation of technology: examining the linear model', Futures, 16, 217-232.

SLOVIC, P., FISCHOFF, B. and LICHTENSTEIN, S. (1980) 'Perceived risk' in SCHWING, R. and ALBERS, W. A., (eds.) Societal Risk Assessment: How Safe is Safe Enough? Plenum Press.

van de PLIGT, J., SPEARS, R. and EISER, R. J. (1984a) 'Public attitudes to nuclear energy', Energy Policy, 12, 302-305.

van de PLIGT, J., EISER, R. J. and SPEARS, R. (1984b) Nuclear Energy: Accidents and Attitudes, Department of Psychology, University of Exeter.

WYNNE, B. (1983) Technology as Cultural Process, Working paper WP-83-118, International Institute for Applied Systems Analysis, Austria.

Chapter Eleven

CHILDREN, RADIATION, CANCER AND THE SELLAFIELD NUCLEAR REPROCESSING PLANT

Alan Craft and Stan Openshaw

Ever since the nuclear bomb was dropped on Japan at the end of the Second World War and the subsequent discovery that many of those who survived the bombs later developed cancer, there has been an almost universal fear of radiation. Any incident or event connected with exposure to radiation generates enormous alarm and concern. Yet if one considers the tiny number of people who are known to have developed cancer due to exposure to radioactivity and compares it to the enormous numbers dying as a result of, for example, accidents or from the effects of smoking tobacco, there is a vastly disproportionate amount of emotional energy spent on the irradiation problem. Perhaps this is because the public has no control over exposure to irradiation and it is a hazard which cannot be seen or sensed. If radioactive discharges were black or turned the sea red then there might not be so much concern! However we are left with the fact that the subject of nuclear energy is a highly emotive one. Indeed what could be more emotive than the popular view of a nuclear fuel reprocessing plant (Sellafield) collecting radioactive waste from all over the world, and reprocessing it - not for any altruistic reason but merely to generate profit which as a by-product eventually results in the apparent death and injury of 'innocent children' from cancer. The general public do not usually seem capable of accurately evaluating risks in life, and emotion plays a far greater part in personal decision than logic. Because of this both government and industry must react to emotion as well as logic. In this chapter we will examine the evidence for the contention that there is a significant link between the nuclear fuel reprocessing facility at Sellafield in West Cumbria and childhood cancer in the Northern

Regions of England, UK, as a case-study of the impacts of nuclear activities on local populations.

INTRODUCTION

In November 1983 Yorkshire Television produced a documentary, 'Windscale: the Nuclear Laundry' which drew attention to the apparent increased levels of cancer among young people along the Cumbrian coast. It related these health effects to the discharge of large amounts of liquid radioactive waste from the Sellafield nuclear fuel reprocessing plant since the early 1950s. There is no doubt that large amounts of radioactive waste have been discharged into the Irish Sea making it one of the most radioactive stretches of water in the world. The Yorkshire Television programme generated considerable public concern and the Government's response to it was to set up an inquiry under the chairmanship of Sir Douglas Black. This resulted in the Black Report (1984). Black reviewed evidence from many sources including data from the Northern Children's Malignant Disease Registry relating to the occurrence of cancer in children not only in Cumbria but throughout the whole of the North of England.

The Report resulted in a tremendous surge of interest in the geographical patterns of cancer incidence, not only around Sellafield but also generally in the country and specifically around other nuclear installations. However, the first report of a cluster of leukaemia cases occurred in 1933, long before any atomic bombs or nuclear industry (Kellett, 1937). Before considering the whole question of the geographical distribution of cancer in children in the Northern Region it is pertinent to pose the question: is there evidence that radiation does cause leukaemia or other cancers in children?

Evidence to support this hypothesis comes from two sources. First there is that from the atomic bomb experience in Japan. It is clear that a proportion of children who are exposed to a single large, sublethal, dose of radiation will develop leukaemia and other cancers at a later date and the evidence for this has been recently reviewed by Beebe (1982). From this experience it has been assumed that there is a continuing straight line relationship between the dose of radiation and the risk of developing cancer and on this assumption have been based the safety standards and limits on exposure to radiation throughout the world. However, it must be made clear that we are dealing in no more than assumptions. The only direct evidence that a single small dose of radiation administered to a child will result in an increased incidence of leukaemia is that seen when babies are irradiated during the first three months of pregnancy. Initially the work of Dr. Alice Stewart produced evidence that there was a twofold increase in the risk of cancer if an X-ray was taken during the first trimester of pregnancy and this has been subsequently confirmed in a study from the United States by Harvey et al., (1985).

There is, however, no evidence that the continuing exposure of children to very low levels of radiation is damaging to their health. The only studies which have reported any evidence of health effects are mainly in adults, and these are in uranium miners, painters of luminous watch dials, and both children and adults given thorium to treat tuberculosis. There has never been a situation where children have been chronically exposed to increased levels of either external or internal radiation over a long period of time.

It is important to make a distinction between the different sorts of radiation to which children are exposed. External radiation is made up primarily of natural background radiation and exposure to medical X-rays, and a very small proportion comes from the radiation legacy from nuclear weapons fallout and the re-entry of nuclear powered satellites. The natural background radiation level varies considerably throughout the United Kingdom but in all areas it is by far the major component of radiation to which children are exposed. A second, and perhaps more important, cause of radiation, especially when one is considering the generation of leukaemia in children, is that produced by the ingestion or inhalation of radionuclides. The most important of these is plutonium which, when ingested, remains in the body for many years, being concentrated in various tissues, including bone. Ingested or inhaled plutonium can come from only two sources: either nuclear weapons fallout or nuclear fuel plant discharges. The amount of plutonium ingested by individuals in different areas of the country is currently under study, but as yet we do not know what levels are likely to be. However, it is apparent that people living in Cumbria are far more likely to have been exposed to plutonium coming from the nuclear fuel reprocessing plant; but equally they may also be at a greater risk exposure to plutonium resulting from nuclear weapons fallout - simply because this is related to rainfall and the west of England has a higher rainfall than many other areas of the country.

The possibility that West Cumbria presents a unique situation in which to study the effects of chronic exposure to lower levels of radiation in children makes it of paramount importance to record accurately all cases of childhood cancer within the area and also to compare these with similar areas which are not exposed to the same apparent radiation hazards. For this reason we have used the Northern Regional Children's Malignant Disease Registry to clarify what would appear, to the casual observer, to be a very clear cut situation of an excess of cases of childhood cancer in Seascale.

Seascale is the nearest village to the Sellafield Nuclear Reprocessing Plant (see Figure 10.1 in Chapter 10 of this volume) and the Yorkshire Television programme claimed that during 30 years, when only 0.45 cases of childhood cancer would have been expected, six cases had in fact occurred; a 14-fold excess. Simple statistical calculation shows this to be highly significant. However, the interpretation of the incidence of rare events in small areas is exceedingly difficult and the available statistical methods can give conflicting answers. These methodological problems are compounded by the difficulties of explaining to a lay audience what the results mean, since they strongly depend, and are conditional, on various statistical concepts with their attendant uncertainties. Yet, both the public and the nuclear industry need a clear and definite answer which is free from all uncertainty, because of the tremendously important social, economic, and political factors involved. There are also many different and powerful vested interests with a stake in the results. They might well be seriously embarrassed if it were to be publicly discovered that either the existing controls on radiation exposure were inadequate or that harmful effects of radiation occurred at much lower exposures than currently thought likely, and that many people's health had consequently been damaged. Alternatively, there are a number of anti-nuclear groups who are equally adamant in their views, and who would be just as angry if it were to be conclusively demonstrated that Sellafield (and by implication all other nuclear installations, since none other come anywhere near releasing such large amounts of radiation), was in fact completely blameless. In this area, any definite conclusion will attract both praise and odium in more or less equal quantities. It will also probably be many years before any really conclusive results will emerge. In the meantime, all that can be done is to review and re-examine the available evidence.

THE BLACK REPORT

In some ways it would appear that the Black Report actually vindicated the fears expressed in the television programme by demonstrating that the incidence of childhood leukaemia in Seascale was far above the national average. It was also accepted that radiation is currently the only known cause of leukaemia in children. Some critics thought that the only possible outcome would be a recommendation that the Sellafield plant should be closed and that a major review of nuclear safety standards would be initiated. Instead, the Black Report concluded that the link between Sellafield and the local cancer rate was not proven and it offered a 'qualified reassurance' to the public about their safety while at the same time recommending an extensive programme of further research. The Government has accepted the Report without reservation and further research is in progress.

The evidence reviewed by the Black Committee is complex and capable of quite contradictory interpretations. For instance, the age-standardised incidence of malignant disease for West Cumbria as a whole is lower than the national average; but the local cancer rate in Seascale village and Millom Rural District is between two and ten times higher than the average leukaemia rate. These findings are based, however, on very small numbers of cancers, since childhood cancers are very rare. When the data were analysed at a 1981 census ward level for the whole of northern England, a number of apparently similar high incidence areas were found outside Cumbria and well away from nuclear installations (Black, 1984; p. 30-32). Additionally, when the data were mapped, no obvious geographical patterns emerged (Craft, Openshaw, Birch, 1984). Figures 11.1 and 11.3 show the distribution of wards with total cancer and leukaemia incidence which may be regarded as being unusually high as

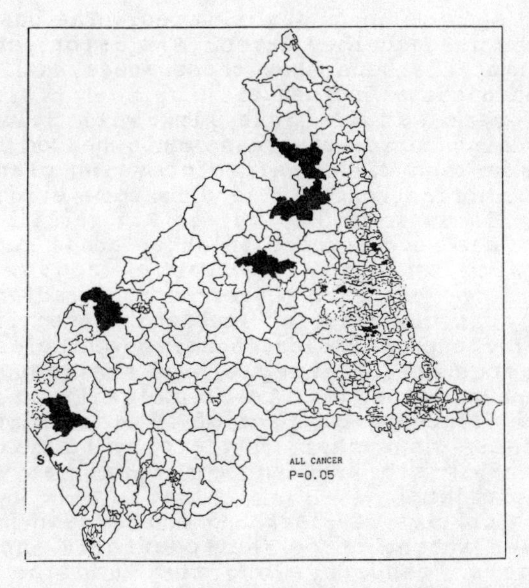

Figure 11.1 Wards with unusually high total childhood cancers; poisson probabilities less than 0.05 (S = see scale)

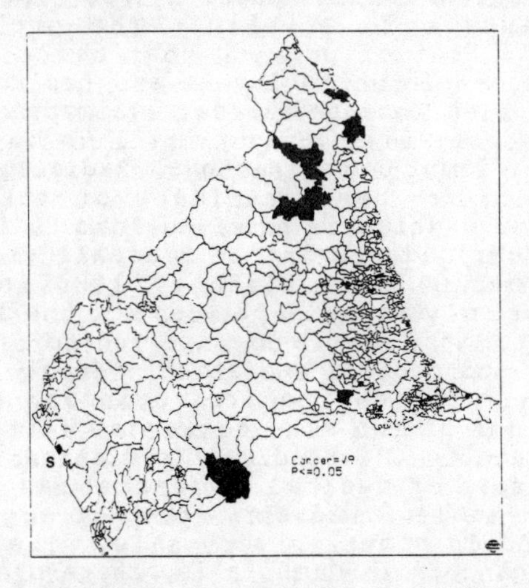

Figure 11.2 Wards with unusually high acute lymphoblastic leukaemias; poisson probabilities less than 0.05

measured by poisson probabilities. The basis for these maps is described later. The important point here is that the maps show those areas with unusually high disease incidence.

If there was a simple link with Sellafield then the high incidence areas would be geographically associated with the reprocessing plant and, or, the Cumbrian coast. There is some evidence to suggest this association, but as Figure 11.2 shows, there are also occurrences in other areas not associated with any known radiation sources. Of course, no one has ever suggested that radiation is the only possible cause of leukaemia; even if it is currently the only proven cause; nevertheless, the map patterns are not simple, and the responsible processes do not appear to be amenable to simple analysis. If there is a set of relationships present in these maps then they are by no means obvious, and herein lies the principal, as yet unsolved, problem.

Some critics of Black and the nuclear industry have drawn attention to the apparently increased incidence of cancers along the Cumbrian coast. This they readily explain as due to the radioactive wastes which are retained in the offshore sediments. These are used to explain the high levels of plutonium found in estuarine areas; for example, the Ravenglass estuary mud has been quoted as having a plutonium content about 27,000 times greater than elsewhere in Britain. The problem, at present, is that of proving that there is a link between these plutonium levels and health effects. It is a problem because current estimates of public radiation doses are far too small to be significant. For example, the National Radiological Protection Board's 'best' estimate of the average radiation dose which is received from Sellafield by under 20 year olds living in Seascale village for the full period from birth to 1980 is able to account for only about 20 per cent of the leukaemia cases which might have been expected to result from background and natural radiation. So it is impossible, given current medical opinion, to assert that even one of the six leukaemias found in Seascale village is the result of radiation.

This sort of medical 'evidence' has been used by the pro-nuclear industry lobby to suggest that the Sellafield plant is very safe. The apparent excess of cancers in Cumbria is, therefore, regarded as purely coincidental, and quite unrelated to proximity to either Sellafield, the Irish Sea, or ionizing radiation. This confidence is reflected in various public statements by the management of the Sellafield plant. For example, they consider that the radioactive discharges could be reduced almost to zero but that there is no sound scientific reason for so doing. The current planned reductions are regarded as necessitated by politics rather than science. They are viewed as unavoidable responses to public pressure, which are not however cost effective since they are thought to prevent, at best, two cancers over the next 10,000 years (at a cost of £120 million each!).

An alternative view is that the public should not in any case expect that the nuclear industry would admit liability for anything. There is far too much at stake. The estimated discharge levels from Sellafield may be incorrect, and could even be gross underestimates. For instance, the earlier discharges associated with Ministry of Defence activities at Sellafield are almost certainly less well established, and the true levels could easily be hidden by national security measures. Additionally, the estimates of the doses received by the public could also be grossly in error; as might be the dose-response assumptions used to compute likely health effects. Presumably, some of the research recommendations made by Black will help to reduce levels of uncertainties in these areas. Nevertheless, one might well expect that if low level radiation is going to have any unexpected health effects, then the evidence will become visible in Cumbria first. Environmental groups talk about 'time bombs' that are long lived, impossible to remove, and perhaps in the long term lethal. But it is too soon to be certain whether or not a real danger exists and to ascertain the likely magnitude of the long term health consequences. In terms of the highly polarised arguments of the pro- and anti-nuclear lobbies, we should either evacuate the area or re-invest the money being spent on pollution control measures in building new hospitals, since these would be certain ways of saving lives.

Neither these arguments nor the actors who make them are rational. It is very difficult to be neutral because of the highly emotive nature of childhood cancer. Accordingly, each new piece of research will be both warmly applauded or rejected, depending on the parties involved and the nature of the findings. The rest of this chapter is a contribution to this debate. It seeks to extend the analyses offered as evidence to the Black Committee of Inquiry by Craft and Openshaw (1984), by broadening the range of cancers considered and examining in more detail certain aspects of the geographical evidence. The chapter describes the data that are studied, reports the results of various spatial analyses designed to uncover and validate any geographical patterns that may be present, and offers general conclusions.

NORTHERN REGION CHILDREN'S MALIGNANT DISEASE REGISTRY

The best source of data for this area on cancer among children is that of the Northern Region Children's Malignant Disease Registry, held in Newcastle. This was established in 1968 with the aim of registering details of all children diagnosed as having cancer before their fifteenth birthday. Registration is primarily through direct notification by clinicians or pathologists, supplemented by cross-referencing with the national Cancer Registration System, and by data from hospital activity analysis. The diagnoses are checked by both pathological review and examination of clinical notes. The data are as complete as they can be and are estimated to cover at least 98 per cent of all cases. The registry provides details of the location, sex, and age of all children who lived in the northern region at the time of diagnosis from 1968 to 1982. Data for 1983 were not complete when the Black Committee was gathering evidence. The locations of children at the time of diagnosis were postcoded, converted to 100m grid-references by a commercial company, and then assigned to 1981 census wards using a computer technique. Estimates of the number of children aged 0 to 15 in 1981 were obtained for the 675 wards in Cumbria, Nothumberland, Tyne and Wear, Durham, and Cleveland; the counties covered by the cancer registry area.

The original evidence given to Black contained a number of caveats about the data which do not appear in the published Report. In general, these data were the best available at short notice in early 1984. There are several limitations which need to be kept in mind; the census counts of children are only relevant for 1981 and thus incidence rates computed for the period 1968 to 1982 may be incorrect; a few cancers were lost due to errors in the post-code grid-reference index held by the post office; the assignment of cancers to census wards will not always be correct; it may be considered desirable to increase the age range of the registry from 0 to 15 to 0 to 25; no information was available about places of birth; and re-analysis at the census enumeration district level was not then possible. These data deficiencies cannot be easily remedied without an extensive clerical exercise, yet because of the small numbers involved and the significance of only one additional cancer in certain areas, it is important that, as far as is practicable, 100 per cent correct data are used for this type of study. These problems were recognised in the Black Report and Recommendation 4 of that Report is a reflection of this need for far greater accuracy and a more detailed re-analysis. The present chapter is, however, restricted to the data that were used by the authors in their submission to the Black Committee in 1984.

The cancer registry records 32 different types of cancer. For convenience, it is necessary to group them together in order to obtain reasonable frequencies. The usual medical groupings would yield the seven variables shown in Table 11.1.

Table 11.1

CANCER VARIABLES

variable	description
1	leukaemia
2	brain cancer
3	kidney cancer
4	neuroblastoma
5	bone cancer
6	lymphomas
7	miscellaneous (i.e. remainder)

Of these some of the leukaemias may be caused by radiation while some lymphomas and bone cancers may possibly also be attributable to radiation. Of the others, cancer of the kidney (Wilms' tumour) is a useful cross-check in that it has been thought to be completely random with similar incidence rates throughout the world, and thus unrelated to environmental factors.

The intention was to compute incidence rates for the 675 wards in the northern region and then examine the cancer profiles of the wards in terms of these seven cancer types. However, childhood cancer is a very rare disease and this causes a number of problems. The Black Report preferred to rank wards by rates per 1,000 children. This has been criticised on the grounds that as most wards only have 0 or 1 cases, all that is reflected in these rates are variations in ward size: this varies by about a factor of 20, from 100 to 2,000 children (Pomiankowski, 1984). On the other hand, the rates are easy to understand. They are a direct measure of average risk defined by the proportion of children registered for a particular type of cancer over the period 1968-1982 when compared with the 1981 child population.

Another measure is the cumulative poisson probability which has long been used by geographers for mapping disease (White, 1971). This gives the probability in a ward of the occurrence of 0, 1, 2, 3 etc. cases, under the assumption that the cases are randomly distributed between wards and come from a population with a constant mean rate of cancer. It is a measure of the chance of finding either the observed cancer rate or even a higher one if the true ward rate were the same as the global mean rate. It has the advantage of standardising for the effects of differences in zone size. The principal disadvantage is that it can easily mislead people when it comes to interpreting the resulting probabilities. Statisticians would caution against putting anything other than a qualified interpretation upon them. However, it is common practice to treat them as a true test of significance.

The two measures are therefore not alternatives but are measuring quite different aspects of the geographical pattern of a cancer (Gardner, 1984). So both rates per 1,000 and poisson probabilities are computed for the seven cancer types in Table 11.1 for each ward.

AREA RANKING

Black Report

The Black Report reproduced parts of two tables from Craft and Openshaw (1984) and then re-ranked the wards by rates per 1,000 instead of the original poisson probabilities. Table 11.2 shows what was published. This has attracted a considerable amount of criticism mainly, it seems, because the Seascale ward ranks sixth in terms of its total cancer rate

---

Table 11.2

RANKING OF WARDS BY TOTAL CANCER AND LYMPHOID MALIGNANCY IN THE NORTHERN REGION

ward rank	number of cancers	population aged 0 to 15	rate per 1,000	poisson probability
all cancers				
1	2	97	20.6	.0128
2	2	133	13.8	.0268
3	2	165	12.1	.0344
4	2	183	10.9	.0415
5	3	281	10.6	.0137
6*	4	411	9.7	.0063
7	6	676	8.8	.0013
8	2	231	8.6	.0626
9	8	953	8.3	.0003
10	5	605	8.2	.0046
lymphoid malignancy				
1	2	144	13.8	.0036
2	1	97	10.3	.0573
3*	4	411	9.7	.0001
4	1	165	6.0	.0955
5	1	172	5.8	.0993
6	1	174	5.7	.1004
7	1	184	5.4	.1059
8	1	189	5.2	.1086
9	1	198	5.0	.1135
10	1	203	4.9	.1162

Source: Craft and Openshaw (1984). Black (1984) Tables 2.18, 2.19

Note: * indicates Seascale ward

and third in terms of lymphoid malignancy rate but it ranks fifth and first when ranked by poisson probabilities. This has been variously interpreted as indicating that the Black Report was biased and that if poisson rankings had been used a very different conclusion would have been obtained. This would be a very naive view because it would require that the entire Report depended on the results reproduced here in Table 11.2. In any case, as Figures 11.1 and 11.2 show, the high incidence wards are not only in proximity to Sellafield even when measured in terms of poisson probabilities.

So far the debate has tended to avoid tackling the underlying statistical problems as to whether or not the observed incidence of leukaemias is anything other than random. If it is random then the observed results are of no significance and proximity to Sellafield is irrelevant. Indeed the apparent clustering of leukaemias in a few wards is a very characteristic feature of rare diseases and might be expected to occur with a purely random generating process. So even with a random distribution of cancers it is inevitable that some cases will occur in areas with small populations, simply because the ratio of total cases to total population is so small and the cases have to occur somewhere. So although the occurrence of four leukaemias in Seascale ward is distinctly odd it could still be a chance occurrence due to random effects.

A closer look at poisson distributions

Unfortunately, further examination of the distribution of cancers by ward is complicated by the uneven population sizes of the census wards. Another problem is that different results might well be obtained if the same data were examined for larger or smaller areas. The only satisfactory approach is to resort to simulation methods and use data for the smallest available areas, which are currently census wards. It is possible to generate an artificial childhood cancer data set that conforms to poisson distribution. The idea here is to employ the same methods of analysis as before but this time applying them to purely synthetic (or artificial) randomly generated data which cannot possibly be related to environmental radiation; and is therefore data which contain no geographic patterns because of its random origins. This is useful because it provides a test of our ability to interpret statistics such as poisson probabilities. The artificial childhood cancer data are generated as follows. There are 635,980 children aged 0 to 15 in the Northern Region. An index is created in the range 1 to 635,980 so that each child is given the ward code in which it is located. Suppose that there are k occurrences of a particular cancer. The random distribution can be obtained by generating k uniformly distributed random integers between 1 and 635,980 and then allocating each to whichever ward it is indexed by.

Table 11.3

RANKING OF WARDS BY ALL LEUKAEMIA AND LYMPHOMAS: BOTH REAL DATA AND A SAMPLE OF 7 SIMULATED DATA SETS

ward	simulated random data from a poisson distribution							real
rank	1	2	3	4	5	6	7	data
1	.003	.0005	.002	.0066	.001	.0001	.002	.0002
2	.004	.002	.006	.007	.002	.005	.008	.004
3	.01	.002	.008	.008	.004	.006	.013	.005
4	.01	.003	.009	.010	.009	.01	.01	.009
5	.01	.024	.01	.01	.01	.01	.01	.01
6	.01	.026	.01	.01	.02	.02	.01	.01
7	.02	.029	.02	.01	.03	.02	.02	.03
8	.02	.031	.03	.01	.03	.02	.03	.03
9	.02	.034	.03	.02	.03	.02	.03	.03
10	.03	.034	.03	.02	.03	.02	.03	.04

Note: the poisson probabilities reported here for the real data differ from those in Table 11.1 where the only leukaemias included are acute lymphoblastics

Table 11.3 reports the results of seven simulations of all leukaemias and lymphomas (variables 1 and 6) with the poisson probabilities for the 10 top ranking wards being listed. The eighth column shows the observed data with Seascale ranked first. It is apparent that there is little obvious difference between the observed data and the randomly generated results. Indeed simulation 6 actually produces lower poisson probabilities than those observed for Seascale. However, this is purely a chance occurrence, it would seem to indicate that the Seascale results in Table 11.2 that have attracted so much attention may also be chance (and thus random) occurrences. Table 11.4 extends the comparison to compare the frequency distributions of the observed and simulated poisson probabilities; once again no real differences appear to exist. The only notable effect concerns the fact that the population sizes for the top few wards with the simulated data are considerably larger

Table 11.4

OBSERVED AND EXPECTED WARD POISSON FREQUENCY DISTRIBUTIONS

	poisson probabilities				
data	less than 0.001	less than 0.01	less than 0.05	less than 0.10	less than 0.15
actual	1	4	13	35	50
simulation 1	0	2	11	31	47
simulation 2	0	2	14	31	54
simulation 3	1	4	12	35	49
simulation 4	0	4	16	29	42
simulation 5	0	3	17	35	51
simulation 6	0	4	11	24	45
simulation 7	1	3	15	29	42

Table 11.5

WARD POPULATIONS FOR TOP RANKING AREAS

ward	simulated random data							real
rank	1	2	3	4	5	6	7	data
1	2680	1902	388	3790	722	1971	1288	411
2	940	1261	1022	183	377	487	1738	144
3	618	1880	1738	1111	891	2262	246	976
4	620	418	1140	1178	1142	2749	1972	609
5	255	858	1203	739	1279	722	280	1207
6	287	1556	287	740	3092	802	780	1353
7	2213	924	1428	1410	947	1463	327	1632
8	1530	384	377	792	947	2197	949	995
9	370	406	1021	1428	395	320	1720	428
10	409	3344	1034	1583	1713	1484	1046	1865

than those observed with the real data; see Table 11.5. At face value these simulation exercises would seem to indicate that ward level cancer distributions similar to those that have been observed can be obtained by chance with a modest degree of regularity. However, there are certain caveats. First, there appear to be qualitative differences at the top of the rankings between the random and the real data. Second, if the years 1983 and 1984 had been included the position of Seascale ward and probably one other in Cumbria might have become far more extreme. Third, the definition of the variables to be ranked may have an appreciable effect even on the existing data. This can be readily tested.

Table 11.6

TOP 10 WARDS RANKED BY ACUTE LYMPHOBLASTIC LEUKAEMIA

ward rank	random data from a poisson distribution							real data
1	.001	.0006	.005	.003	.002	.0004	.003	.00003
2	.011	.005	.013	.008	.017	.008	.005	.001
3	.002	.017	.015	.008	.024	.012	.010	.006
4	.028	.020	.018	.015	.032	.016	.017	.014
5	.030	.022	.022	.015	.035	.024	.024	.017
6	.031	.022	.024	.017	.041	.027	.026	.020
7	.o32	.022	.027	.020	.042	.027	.027	.022
8	.034	.026	.030	.022	.049	.031	.028	.022
9	.037	.027	.031	.027	.056	.038	.048	.025
10	.040	.027	.033	.029	.057	.039	.056	.027

Table 11.6 reports the results of ranking only acute lymphoblastic leukaemia of which the four Seascale cancers consist. The situation concerning the top two or three wards is now changed. The poisson probability for Seascale ward becomes much smaller and may now be significantly different for the first time from that which can be produced at random. A significance test can be used to test this hypothesis. A sample of 99 different random simulations of acute lymphoblastic leukaemias was generated and their poisson probabilities were computed. The observed Seascale value can be ranked with the 99 random values and an approximate

level of significance obtained. The null hypothesis of no significant difference between the distributions can now be rejected at the 5 per cent level. The second, third and fourth ranked wards are not significantly different at either the 5 or 10 per cent levels. This would suggest that the leukaemia rate for Seascale is the only real abnormality uncovered by ranking wards. It will be interesting to see whether or not subsequent occurrences of the disease will confirm this finding. Preliminary analysis of the 1983 and 1984 data suggests that this might be the case.

MULTIVARIATE CLASSIFICATION OF WARDS BY THEIR CANCER PROFILES

So far attention has been focused on univariate analyses and there is a danger that more subtle multivariate patterns may have been missed. A useful data exploratory technique to help discover any such patterns is that of numerical taxonomy. A multivariate classification technique can be used to group wards together on the basis of similarities defined in terms of their childhood cancer profiles. If there is any clustering of cancers within the northern region then this classification approach will help to identify them, when the wards belonging to each profile are mapped. The resulting map patterns can then be used to test the Yorkshire TV hypothesis that the Cumbrian coast is a region of increased cancer rates and also to indicate whether any other parts of the Northern Region have a similar pattern of childhood cancers and thus, whether the Cumbrian wards are in any way unusual.

A 10 group classification

A number of computer generated classifications of the 675 wards were made for the variables defined in Table 11.1 using the technique described in Openshaw (1982). Each variable was represented in two forms; once as a rate and once as a poisson probability. A 10 group classification was used to illustrate this process of data exploration. Table 11.7 provides a list of the cancer area types that were identified and readily demonstrates the utility of the apparent method as a means of obtaining summaries of multivariate data.

Table 11.7

AREA TYPES IN A 10 GROUP CLASSIFICATION

group	wards	description
1	20	very high lymphoma rate (16* average)
2	54	high brain cancer rate (6*)
3	17	high leukaemia (10*)
4	24	high miscellaneous cancer rate (11*)
5	12	very high bone cancer rate (28*)
6	17	very high kidney cancer rate (17*)
7	189	slightly higher than average cancer rates
8	11	very high neuroblastoma rates (30*)
9	140	average cancer rates
10	91	lower than average cancer rates

There are six high cancer incidence groups (1-5,8) and the location of these areas is shown in Figure 11.3. Despite this discovery, there does not appear to be any obvious geographical pattern in these maps. Moreover, group 6 (high kidney cancer rate) shows a clustered pattern similar to several others, but this particular cancer is thought to be completely random in occurrence, so here at least the map pattern evidence is completely misleading. It is possible then that the distribution of areas with different types of cancer incidence is merely, or largely, a random one. This belief is supported by the failure of the different types of cancer to be related and thus form overlapping distributions; this follows from the fact that each group is characterised only by one dominant cancer, suggesting that there are no geographical relationships between them. This may or may not be an important finding. It is difficult to interpret because there is no other work of a similar nature to compare it with. The prospect that the classification results are largely random can be investigated using the artificial randomly generated data. Table 11.8 shows the labels given to <u>one particular</u> random set of data that was classified and some of the groups are mapped in Figure 11.4. Now it would be expected that completely uninterpretable results would be

Table 11.8

AREA TYPES IN A 10 GROUP CLASSIFICATION: RANDOM DATA

group	wards	description
1	126	average cancer rates
2	169	slightly above average rates
3	30	high miscellaneous rates (9* average)
4	49	high leukaemia rates (5*)
5	22	high brain rates (5*) and very high lymphomas (13*)
6	8	very high neuroblastoma rates (38*)
7	30	very high bone cancer rate (18*)
8	11	very high kidney cancer rate (30*)
9	125	average cancer rates
10	105	low cancer rates

obtained from classifying random data. Unfortunately the group labels identified in Table 11.8 must be considered as quite interpretable and their apparent general similarity to Table 11.7 suggests that random events rather than Sellafield are responsible. A final indication of this is to count the number of wards located in high incidence areas, in the hope that the random data classification may have produced a far less geographically concentrated distribution. Sadly, it has not. There are 150 high cancer incidence wards in the random data classification compared with 155 in the real data classification, both with 10 groups. So at best any non-random effects are slight when measured against data for the entire Northern Region.

A revised classification

It would appear that at best any map associations are weak and difficult to find. It may be that the definition of the variables in Table 11.1 is critical because the occurrence of only one or two extra cases in an area is sufficient to completely alter a ward's profile. The leukaemia variable may be poorly represented in the 10 group classification merely because it is poorly correlated with any of the other variables. This situation can be altered by including additional variables in the

Figure 11.3 Groups of wards with high cancer incidence

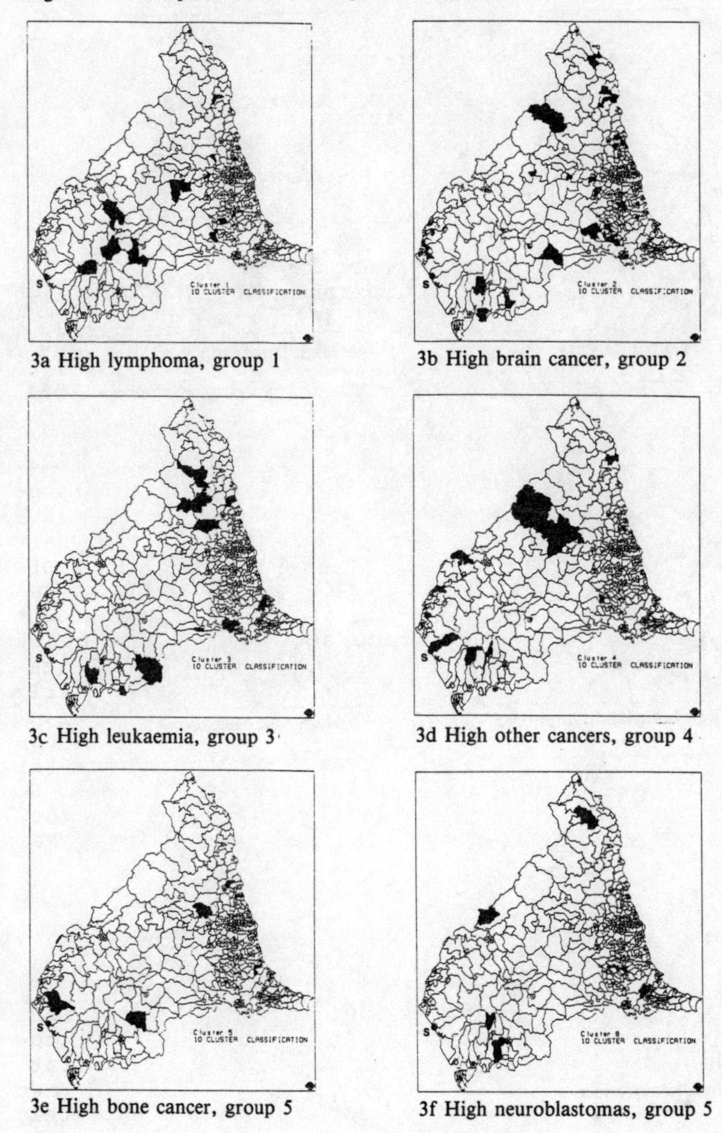

3a High lymphoma, group 1

3b High brain cancer, group 2

3c High leukaemia, group 3

3d High other cancers, group 4

3e High bone cancer, group 5

3f High neuroblastomas, group 5

Figure 11.4 Groups of wards with high cancer rates based on random data

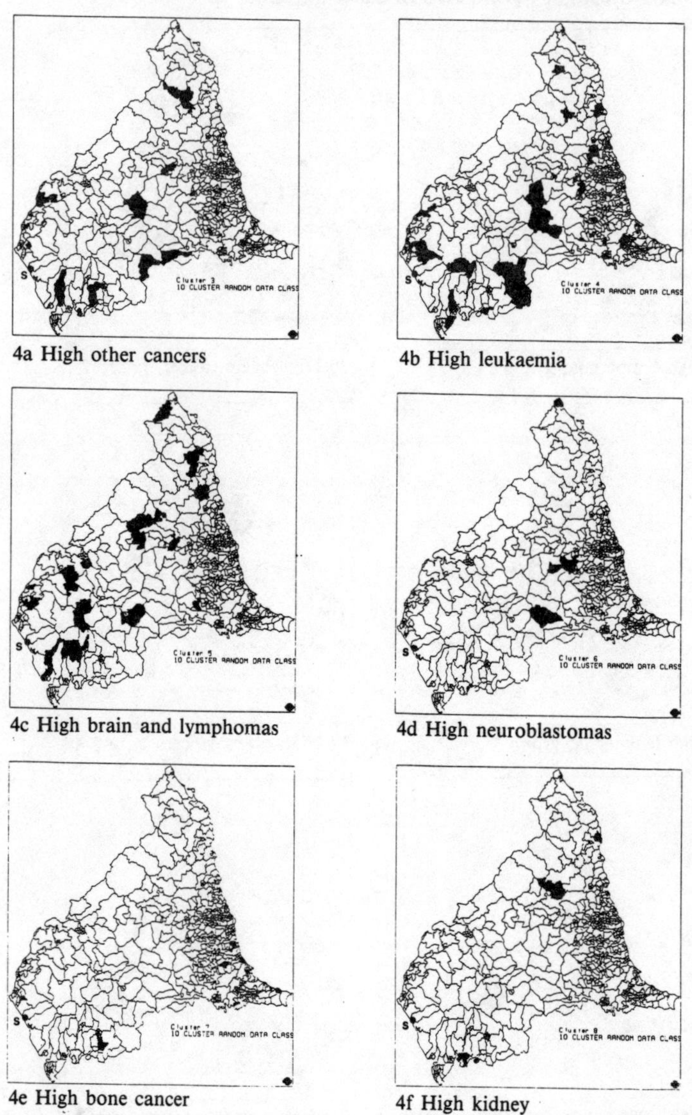

4a High other cancers

4b High leukaemia

4c High brain and lymphomas

4d High neuroblastomas

4e High bone cancer

4f High kidney

analysis which repeat certain aspects of the variables already used. Four additional variables are defined as: acute lymphoblastic leukaemia, acute lymphoblastic leukaemia with non-Hodgkin's lymphoma, all lymphoid malignancy, and all childhood cancers. The variables are again represented as both rates and as poisson probabilities. The hope is expressed that the deliberate inclusion of practically double-counted variables will improve the distinctiveness of the high cancer areas which are of interest. Table 11.9 shows the labels given to the preferred 15 group classification and the distribution of selected groups is shown in Figure 11.5. There are now 140 wards in the eight high cancer groups. Seascale is in group 3 along with Elsdon and Whittingham, both of which are in Alnwick District, Northumberland.

Table 11.9

AREA TYPES IN A 15 GROUP CLASSIFICATION

group	wards	description
1	57	low cancer rates
2	140	slightly higher than average rates
3	19	very high miscellaneous cancers (12* average) and high all cancers (3*)
4	117	very low leukaemia rates
5	12	very high bone cancer rates (28*)
6	10	very high lymphomas (22*), high malignant lymphomas, brain cancer, and all cancer rates
7	3	very high neuroblastoma rate (56*) and high other cancers
8	34	high brain cancer rate (7*)
9	44	very low cancer rates
10	15	very high kidney cancer rate (25*)
11	3	very high leukaemia (21*), brain (8*), and lymphoid malignancies (25*)
12	22	low lymphoid malignancies
13	115	lower than average cancer rates
14	63	average cancer rates
15	21	very low cancer rates

Once again a random data classification showed a similar distribution, although this time the number of wards included in the eight high cancer areas

Figure 11.5 Revised classification of wards; areas of high cancer incidence

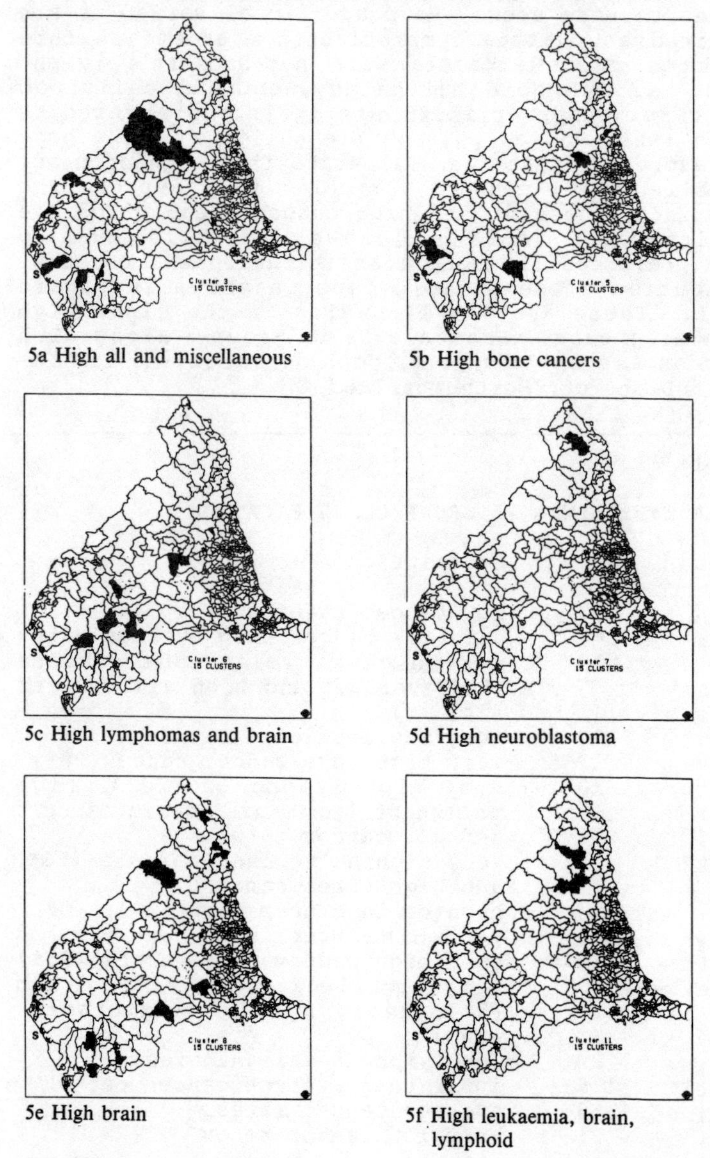

5a High all and miscellaneous

5b High bone cancers

5c High lymphomas and brain

5d High neuroblastoma

5e High brain

5f High leukaemia, brain, lymphoid

totalled 157. Another difference was that most of the clusters consisted of single cancer types but with the real data there are now a number of clusters which were characterised by a mix of more than one cancer type. The geographical map patterns remained, however, difficult to interpret.

Time subdivisions

A final aspect worthy of investigation concerns the possible existence of spatial patterns over time. The data can be disaggregated into three quinquennia: 1968-72, 1973-77, and 1978-82. The breakpoints also coincide with the three different discharge levels from Sellafield which can be characterised as: low, high and moderate. If Sellafield were having an effect, then it can be hypothesised that the excess rates might be first noticed in period one, and will expand in periods two and three. However, none of the cluster analysis results showed any indication of temporal trends although analyses were difficult because of the small numbers involved. The solution must be to extend the geographical domain of the study region so as to increase the number of cancers that can be examined.

This negative result is also reflected in the plots of cancer and Sellafield discharge time series given by Black (1984). There is no relationship, either direct or lagged. For example, the levels of liquid discharge did not start to become large until after 1968 whereas 11 of the 28 leukaemias in Millom RD occurred before that date (Black, 1984, p. 19). Moreover, the 1972-6 peak discharge levels seem to have had no effect so far on cancer incidence. This might be because the BNFL discharge figures are inaccurate or because the distribution of discharges was concentrated rather than being spread out over a whole year and the extent of the temporal concentrations changed. It is also an area where the current lack of medical knowledge relating to the lag time between exposure and generation of a particular cancer prevents more specific and relevant hypotheses being tested.

Conclusions

Considerable publicity followed the publication of

the Black Report and there were many suggestions that the Report represented a 'whitewash' or cover-up. Yet, as can be seen, the geographical evidence based on the 1968-82 data for what may be termed a 'Sellafield effect' on Cumbrian cancer rates is not at all convincing. Nonetheless, the cancer rate in Seascale ward does appear to be exceptionally high - the full nature of it may well be clarified only by subsequent research. Other causative effects may also be present but if they exist they will emerge only when the data base is enlarged. It may also be important to try to include other variables which serve to reduce localised variation; for example, occupational mobility. Additionally, if more were known about the cause of leukaemia then it might be possible to ask more relevant questions and discover additional patterns that are at present not obvious because we really do not know what precisely to look for. The application of computer automated search methods may be of some assistance here, in the future.

The ranking studies and the classifications suggest that, perhaps, the pattern of cancers, or at least certain types of cancer, is more geographically concentrated than might have been expected from a purely random process. This clearly requires further investigation using higher resolution and more accurate data. Other possible geographical hypotheses also come to mind that may 'explain' the non-Cumbrian leukaemia clusters; for example, proximity to microwave installations run by the military, nearby chemical industries, etc. There may well be more than one form of pollution trigger but without a significant extension to the temporal and geographical domain of the study it will be impossible to identify them. The same applies to other questions, such as what proportion of all children with leukaemia in the country have lived in Cumbria, or who had been there on holiday, and these cannot be answered without nationwide coverage. The subject is further complicated by the possibility that individual susceptibility to leukaemia may vary and require different triggers. It is also likely that there is more than one generating process which is responsible for the spatial patterns, and that existing tools are not sufficiently sophisticated to untangle the resulting complexities. The only solution to these problems at present is to enlarge the amount of data available for analysis in the hope that more definitive conclusions may emerge at some future date.

In offering these conclusions we state that we are connected with neither the nuclear industry nor anti-nuclear organisations. This is relevant because both organisations regard science and statistical analysis as a form of public relations exercise. It is important to be realistic. One reason for the high level of media interest is the apparently high-handed manner by which the nuclear industry dismisses public fears. If it was seen to be more honest and more open, and it was perhaps admitted that maybe it could have possibly been responsible for a small number of deaths (and it does not know which ones) but pointed out that it had in fact an excellent record for a major industry - not without some public benefits - then it would be quite likely that media attention would diminish and public acceptance of assurances of safety might increase. Instead it argues that no proven health links exist, and it thereby exploits the inherent uncertainty and random components that inevitably characterise all cancers. This policy is short-sighted and in the longer term it may fail. On the other hand, the strategy employed by the anti-nuclear groups and the media is also illogical and often without any scientific foundation. To them every cancer is a result of criminally high levels of radioactive discharge, every cluster of leukaemia is radiation induced; and they support their claims with dubious statistics. These claims will invariably be strengthened by the discovery of well above background radiation fields around many nuclear installations. Ideally, they should not exist but they certainly do; even if, as far as it is possible to determine, the excess radiation doses are small. What is missing is any critical link with health effects, and if indeed they do exist it may still be 50 years before they are established. Yet even if they are, the number of children with leukaemia in the ward nearest the reprocessing plant will still be considerably less than the number killed in the same period in the same ward by road accidents. On the other hand, it is also impossible at present to determine what proportion of all children with leukaemia have ever spent a significant part of their lives in Cumbria. It is important to keep things in perspective. Given the current evidence, neither side will be able to prove their case. Sellafield can be proven neither innocent nor guilty. The best outcome for both sides would be a greater political acceptance of:

(i) the need to minimise cumulative public exposure to radiation just in case expert opinion is wrong, irrespective of short-term costs;
(ii) the need to establish urgently an extensive and accurate public dose monitoring system;
(iii) to fund basic research into low level radiation effects and appropriate data-analysis systems.

Viewed in this context, the Black Report has been a useful development that should result, eventually, in a better understanding of the problem; indeed the Report is the best that could have been hoped for, given the facts at its disposal.

REFERENCES

BEEBE, G. W. (1982) 'Ionising radiation and health', American Scientist, 70, 35-44.

BLACK, Sir D. (1984) Investigation of the Possible Increased Incidence of Cancer in West Cumbria: Report of the Independent Advisory Group, London: HMSO.

CRAFT, A. W. and OPENSHAW, S. (1984) 'Childhood cancer in the northern region 1968-1982: a geographical analysis, preliminary report', personal communication to Sir Douglas Black, document SDB555/H13 (mimeo).

CRAFT, A. W., OPENSHAW, S., BIRCH, J. M. (1984) 'Apparent clusters of childhood lymphoid malignancy in northern England', The Lancet, 14 July, 96-97.

GARDNER, M. (1984) 'Seascale hazard', Nature, 312,10.

HARVEY, E. B., BOICE, J. D., HONEYMAN, M., FLANNERY, J. T. (1985) 'Prenatal X-ray exposure and childhood cancer in twins', New England Journal of Medicine, 312, 541-545.

KELLETT, C. E. (1937) 'Acute myeloid leukaemia in one of identical twins', Archives of Disease in Childhood 12, 239-252.

OPENSHAW, S. (1982) A Portable Suite of FORTRAN IV Programs (CCP) For Classifying Census Data for Districts and Counties: an Introduction and User Guide, CURDS Report, Newcastle University.

POMIANKOWSKI, A. (1984) 'Cancer incidence at Sellafield', _Nature_, 311, 190.

WHITE, R. R. (1971) 'Probability maps of leukaemia mortalities in England and Wales', in McGLASHAN, N. (ed.) _Readings in Medical Geography_, London: Methuen.

Chapter Twelve

EVACUATION DECISION-MAKING AT THREE MILE ISLAND

Donald J. Zeigler and James H. Johnson, Jnr.

Set into motion by a series of mechanical failures and human errors, the near-catastrophic accident at the Three Mile Island (TMI) nuclear generating station near Harrisburg, Pennsylvania, on 28 March 1979, evoked an unanticipated and unprecedented level of spontaneous evacuation among area residents in response to the threat of an impending nuclear disaster (Figure 12.1). Even more alarming than the intial releases of radioactive steam from the reactor building were the inability of the utility company to bring the reactor back under control and the materialisation of a 'hydrogen bubble' in the reactor vessel. The uncertainty over whether an explosion could be expected if the oxygen content of the bubble increased, the acknowledged potential for a major disaster, the release of several additional puffs of radioactive steam two days after the intial accident, and the inherent fear of the radiation hazard all pointed to an evacuation of the local area. Finally, the Governor of Pennsylvania released the following statement which was deliberately worded to be as reassuring as possible:

> Based on advice of the chairman of the NRC [Nuclear Regulatory Commission], and in the interests of taking every precaution, I am advising those who may be particularly susceptible to the effects of radiation - that is, pregnant women and pre-school children - to leave the area within a 5-mile radius of the Three Mile Island facility until further notice. We have also ordered the closing of any schools within this area. I repeat that this and other contingency measures are based on my belief that an excess of caution is best.

Figure 12.1 THE THREE MILE ISLAND VICINITY OF SOUTH-CENTRAL PENNSYLVANIA

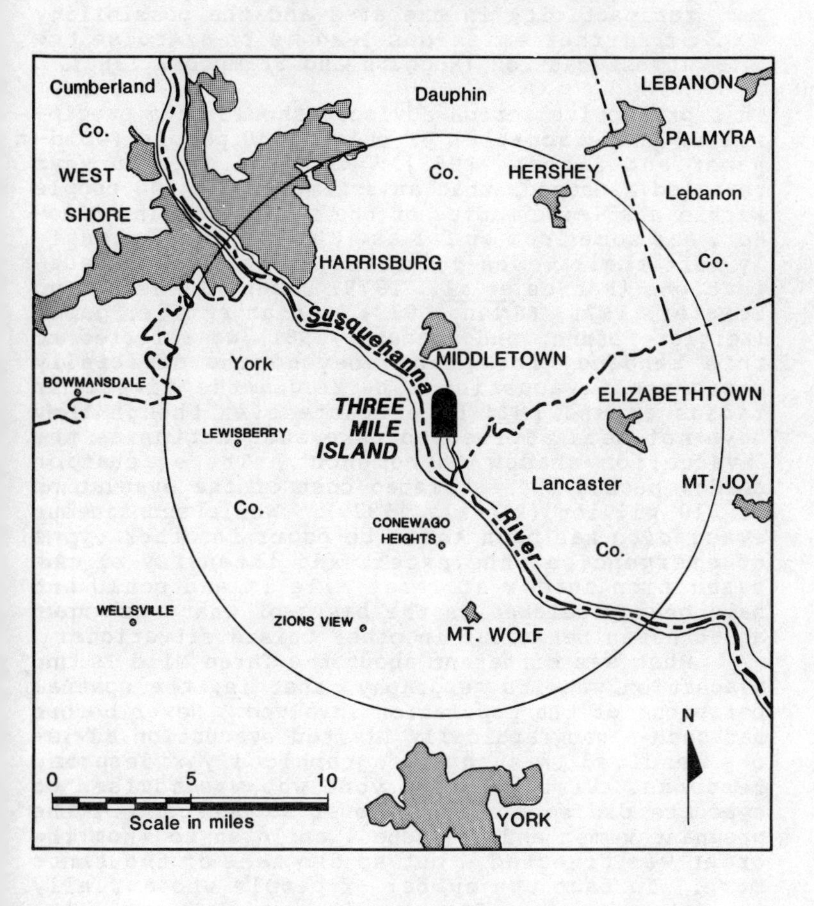

> Current readings are no higher than they were yesterday. However, the continued presence of radioactivity in the area and the possibility of further emissions lead me to exercise the utmost caution (Rogovin and Frampton, 1980).

This protective action advisory should have precipitated the evacuation of only 3,400 people (Goldhaber and Lehman, 1983). Post-accident surveys revealed, instead, that an estimated 200,000 people within a 25-mile radius of the malfunctioning reactor, and some from as far as 40 miles away, actually left their homes for safer, more distant destinations (Barnes et al., 1979; Brunn, Johnson, and Zeigler, 1979; Flynn, 1979). In an earlier paper (Zeigler, Brunn, and Johnson, 1981) we referred to this tendency for people beyond the officially designated evacuation zone (e.g., the five mile radius around TMI) to evacuate even though they have not been advised to take such action as the 'evacuation shadow phenomenon'. The evacuation shadow pushed the estimated cost of the evacuation to $10 million (Plosila, 1979). While spontaneous evacuation has been known to occur in other types of emergencies, the extent and intensity of the evacuation shadow at Three Mile Island could not have been predicted on the basis of what was known about human behaviour in other crisis situations.

What was different about the Three Mile Island evacuation was its geography, that is, the spatial behaviour of the population involved. Never before had such a geographically limited evacuation advisory resulted in such a geographically widespread response. Virtually everyone who was advised to evacuate did so - slightly over 90 per cent of the pregnant women and pre-school children to whom the order was directed - but so did tens of thousands more. In fact the number of people who actually evacuated was over 50 times the number to whom the evacuation order applied. Furthermore, those who left put a median distance of 100 miles (Flynn, 1979) between themselves and the threatening reactor, the longest-distance evacuation on record. While the primary public evacuation centre was established only 10 miles away from TMI, half of all evacuees fled more than ten times that distance.

In evacuation from natural disasters, people have been reluctant to move until they were absolutely convinced that the threat was both real and imminent. Given the results of natural disaster studies, under-response to official evacuation warnings is expected to typify 10 to 50 per cent of the population. Traditionally, therefore, a serious problem has been to devise strategies to encourage everyone to comply with the evacuation order. Furthermore, in the 500 or so natural and technological disasters studied by Hans and Sell (1974), the average evacuation distance was only 13 miles. Even the threat of a chlorine gas cloud in Mississauga, Ontario, in 1979 was not enough to precipitate a long-distance move: one third of the evacuees had to be advised to move a second time in order to avoid the hazard (Liverman and Wilson, 1981). The gulf between the spatial behaviour of evacuees from the nuclear disaster at Three Mile Island and the spatial behaviour of evacuees from other disasters with which we have more experience is both remarkable and significant for the planning process. Where evacuees originate and where they go are fundamental to emergency response planning. Why these spatial behaviours characterise nuclear evacuations is an equally important issue in trying to plan for the aftershocks of future nuclear emergencies.

One of the important conclusions to be drawn from the Three Mile Island experience is that nuclear accidents in many important respects are not equivalent to natural and other technological emergencies. Because people perceive the radiation hazard differently (Slovic, Fischhoff, and Lichtenstein, 1979 and 1980), because they cannot see or sense the hazard agent itself, because they fear the carcinogenic and transgenerational consequences of radiation exposure, and because they have no way of evaluating the veracity of official information, people respond to nuclear accidents in a way that betrays this fear. Perry (1981), for instance, found that only 20 per cent of three flood-threatened populations rated the threat to them as severe, quite in contrast to 50 per cent of the Three Mile Island population. In spite of these documented perceptual and behavioural differences, there is in both the United States and the United Kingdom an official policy of regarding nuclear disasters as equivalent to other disasters, and therefore manageable by the same techniques. In the US there has even been a recent push, led by the American Nuclear Society, to downgrade the likelihood of a nuclear power plant accident. However, after examining the evidence as recently as 1985 the American Physical Society stated that it was not yet possible to conclude that the maximum potential radioactivity release from a severe nuclear accident was much smaller than previously thought. Even if future estimates of worst-case disasters are reduced, however, how is the nuclear establishment going to convey this to the public in terms that they will find persuasive? Given the chaos in the ad hoc planning process that accompanied the accident at Three Mile Island in 1979, it is unlikely that the public will support turning back the clock on emergency planning by acquiescing to a reduction in the size of the evacuation planning zone or to a policy of locating new nuclear power stations closer to urban populations, both of which have been recommended by the nuclear industry. It was the President's Commission (Kemeny _et al._, 1979, 38), after all, which criticised the NRC for having over-confidence in reactor safeguards and for its desire to avoid raising public concern about nuclear power safety.

A number of official inquiries were commissioned by the State and Federal governments in the months after the accident at TMI (Chenault, Hilbert, and Reichlin, 1979; Kemeny _et al._, 1979; Plosila, 1979; Rogovin and Frampton, 1980). On the basis of their findings and recommendations, especially the President's Commission on the Accident at TMI, revised and upgraded emergency preparedness and response regulations were issued jointly by the Nuclear Regulatory Commission (NRC) and the Federal Emergency Management Agency (FEMA) in 1980 (NRC/FEMA, 1980). These new regulations expanded the evacuation planning zone to ten miles; required approved emergency response plans as prerequisites for issuing a licence; and directed nuclear utilities and relevant local, State, and Federal agencies to undertake a public education programme and, in the event of a nuclear accident, provide the public with clear, concise, and accurate emergency information. The revised regulations assume that educating the public and providing timely information during a crisis will minimise, if not eliminate, the propensity of people to evacuate spontaneously. We have argued in several recent papers that this is an unrealistic planning assumption and that plans developed according to the revised Federal guidelines are very likely to be inadequate and ineffective should a nuclear reactor emergency occur in the future (Johnson and Zeigler, 1983 and 1984; Zeigler and Johnson, 1984; Johnson, 1984 and 1985 a, b). We believe that the NRC and FEMA, in attempting to revise and update the nuclear emergency planning regulations, lacked a clear understanding of the importance of spontaneous evacuation and the full range of factors that influenced emergency decision-making during the Three Mile Island accident.

In this chapter we report on the results of an empirical test of a causal model of emergency evacuation decision-making. We have developed this model to assist emergency management officials in their attempt to understand public behaviour during the TMI crisis. We have conducted one previous test of the proposed model, using survey data obtained in interviews with a representative sample of households around a commercial nuclear power plant on Long Island in New York (Johnson, 1985b). The results of that test strongly support the hypotheses developed in this chapter, but the New York data were based on intended rather than actual evacuation behaviour. There are a number of sound reasons to believe that people will act out their behavioural intentions in the event of a nuclear reactor emergency (Ajzen and Fishbein, 1980). Most importantly, the actual behaviour of south-central Pennsylvania residents during the TMI crisis closely parallels the intended behaviour of the New York sample. Nevertheless, the reliability of our model as a predictive tool will remain unknown unless another accident occurs. The results of our analyses support the behavioural-based approach to radiological emergency planning which we have advocated elsewhere. This approach is based on the need to take advantage of the natural protective inclinations of the people rather than trying to control an evacuation by sponsoring public education campaigns and rigidly planning emergency information dissemination. In addition, we feel that the experience at Three Mile Island provides useful insights for other countries' planning efforts. For that reason we have included a brief critique of emergency plans for the Sizewell reactor site in the United Kingdom.

A model of evacuation decision-making

Our model of emergency evacuation decision-making for nuclear accidents is depicted in Figure 12.2. Critical in its development were the findings of behavioural surveys conducted in the aftermath of TMI and work about public perceptions, beliefs, and attitudes to the use of nuclear power.

Figure 12.2: A GENERALISED MODEL OF EVACUATION DECISION-MAKING FOR A RADIOLOGICAL EMERGENCY AT A NUCLEAR POWER STATION

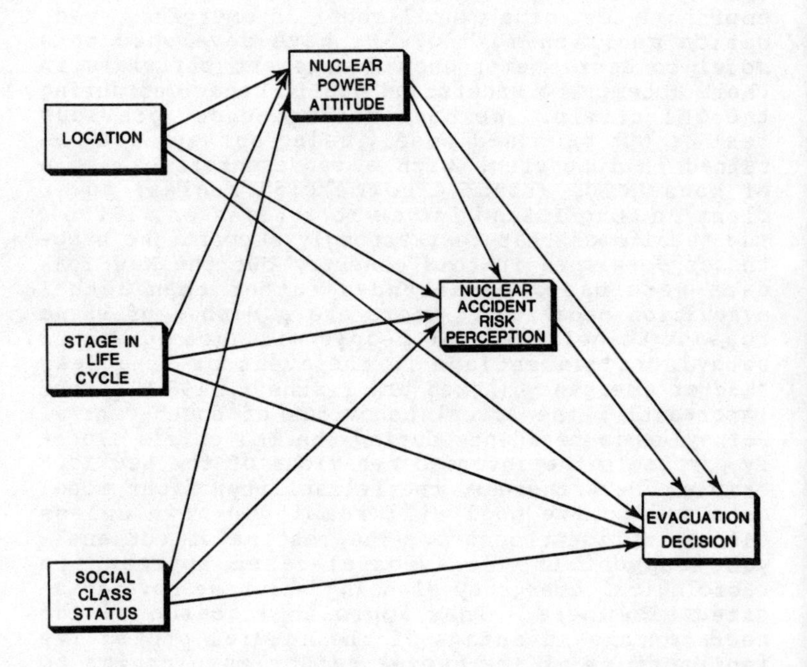

The proposed model comprises three sets of variables over which the respondent has no immediate control (location, stage of the life-cycle, and social status) and three sets of variables which reflect the respondent's evaluation of the nuclear hazard (attitude toward nuclear power, accident risk perception, and the evacuation decision). Potential interactions are shown as pathways, depicted by arrows which show the directions of the suspected relationships. Location, for instance, is expected to influence the evacuation decision directly and indirectly through the attitude and risk perception variables. The structure of the proposed model revolves around five major research hypotheses which may be summarised as follows:

> Hypothesis 1: Where a resident lives in relation to the nuclear power station will directly influence the evacuation decision.

A wide range of locational factors, including proximity to the reactor, physical features in the local environment, local land use patterns, major transport routes, and prevailing wind direction, will influence evacuation decision-making in a radiological emergency. However, post-accident studies at TMI suggest that the two fundamental locational variables which capture the effect of local geography are distance and direction of the home from the plant. The influence of direction on the evacuation response will depend on local characteristics of each site, but the most fundamental influence seems to be wind direction.

> Hypothesis 2: The household's stage in the life-cycle will directly influence the evacuation decision.

Because of the perceived long-term and intergenerational effects of exposure to radiation on the foetus and young children, we believe that families in their child-bearing years (who may or may not already have young children in the home) will be more likely to evacuate than older families without children.

> Hypothesis 3: The head-of-household's social status will directly influence the evacuation decision.

Based in part on the results of referenda voting behaviour (Benedict et al., 1980; Bowman and Fishbein, 1978; Webber, 1982) and in part on the correlates of evacuation from TMI we postulate that higher status households are more likely to evacuate than lower status households.

> Hypothesis 4: The head-of-household's pre-accident attitude toward nuclear power will directly influence the evacuation decision.

Survey research in the US has shown that the variable 'concern about nuclear power safety' is the strongest discriminator between those who oppose and those who favour the use of nuclear power. 'Pro-nuclear individuals believed most in the safety of nuclear power plants, while anti-nuclear individuals believed that nuclear power plants are not so safe, or are dangerous' (Melber et al., 1977, 281). In the light of this and related studies (Firebaugh, 1982; Inglehart, 1984; Otway, 1977; Otway and Fishbein, 1976), we postulate that those who oppose using nuclear power to generate electricity will be more inclined than those who favour the use to evacuate in a radiological emergency.

> Hypothesis 5: The head-of-household's perception of the risks associated with a nuclear reactor accident will directly influence the evacuation decision.

Previous research has shown that the ability of a potentially affected population to confirm official warnings through direct sensory evidence (e.g., the winds associated with a hurricane) is a critical determinant of evacuation behaviour in natural disasters (Perry, Lindell, and Greene, 1981). In a nuclear emergency, however, the disaster agent - ionising radiation - will be invisible and otherwise insensible (except in large enough doses to induce radiation sickness), and therefore, there will be no environmental cues (Zeigler, Johnson, and Brunn, 1983, 81-83). This means that the public will be forced to rely solely on secondary and tertiary sources of information about the emergency. Given nuclear technology's inherent complexity and the low level of public confidence in the nuclear establishment, whatever official information is released during a nuclear accident is likely to be both confusing and distrusted (Media In-

Table 12.1 CONCEPTS, VARIABLE AND SPECIFIC MEASURES

Concepts and Variables	Description
A. Locational	
1. Perceived distance (x_1)	Exactly how many miles do you live from the Three Mile Island Nuclear Power Plant? Exact mileage recorded.
2. Direction (x_2)	Coded into four categories: (1) East (2) West (3) North (4) South.
B. Stage-in-family life cycle	
3. Age of household head (x_3)	Absolute age recorded.
4. Marital status (x_4)	Coded into 5 categories: (1) Single (2) Married (3) Divorced (4) Separated (5) Widowed.
5. Young children in the home (x_5)	Absolute number recorded.
6. Any one pregnant in the home (x_6)	Coded into two categories (1) yes (0) no.
C. Social Class Status	
7. Years of school completed (x_7)	Absolute number recorded.
D. Psychological	
8. General attitude toward nuclear power (x_8)	How do the advantages of nuclear power in general compare to its disadvantages? Coded into 5 categories: (1) advantages much greater than disadvantages (2) advantages somewhat greater than disadvantages (3) advantages and disadvantages about the same (4) advantages somewhat less than disadvantages (5) advantages much less than disadvantages.
9. Attitude to TMI plant (x_9)	Before the accident did you think the advantages of having the TMI nuclear station in your area were greater than, less than or about the same as the disadvantages? Coded in the same 5 categories as in Number 8 above.
10. TMI accident risk (x_{10})	How serious a threat did you feel the TMI nuclear station was for you and your family's safety at the time of the accident? Coded into four categories: (1) Very serious (2) serious (3) somewhat of a threat (4) no threat.
E. Behaviour	
11. Evacuation decision (x_{11})	Did any member of your household evacuate during the 2 weeks of the TMI incident? Coded into two categories (1) yes (2) no.

stitute, 1979 and 1980a, b). Because of all these factors, those heads of household who perceive themselves and their families to be at risk of exposure will be more likely to evacuate than those who perceive that they are not in danger.

In addition to the foregoing direct influences, we also specify in the model a number of indirect influences on the evacuation decision. These are shown in Figure 12.2 to work through the variables which represent nuclear power attitude and nuclear accident risk perception. For instance, a resident's pre-accident attitude toward nuclear power is thought to influence indirectly the evacuation decision by influencing perceptions of nuclear power risk. Those who oppose the use of nuclear power to generate electricity will be more likely to perceive that they are at risk of an accident than those who favour its use. Hence, for example, our proposed model of nuclear reactor emergency evacuation decision-making incorporates both attributes of location and characteristics of individuals' social status and attitudinal position. In the section that follows we apply the model in a case study of evacuation behaviour during the TMI reactor crisis.

Data and methods

Data used in this test of the proposed model were taken from the Nuclear Regulatory Commission-sponsored *Three Mile Island Telephone Survey* (Flynn, 1979). Designed to assess the social, psychological, and economic impacts of the TMI accident, including the extent of evacuation, the NRC survey was administered approximately three months after the crisis to a stratified random sample of 1,500 households residing within a 55-mile radius of the site. Eleven variables were selected from the NRC survey for the purpose of this analysis. Along with the specific questions included in the interview schedule, these variables are listed in Table 12.1. Each respondent's location, stage in the life-cycle, and social status have been associated with one or more variables drawn from the survey. In addition, three psychological variables have been chosen to describe each respondent's attitude toward nuclear power and perception of nuclear accident risk. Attitude toward nuclear power in general and the TMI plant in particular are undoubtedly highly correlated; however, it is possible for an individual to support the use of nuclear power in general but to oppose a plant which has been constructed close to home or which has been poorly managed. Finally, the survey data make it possible to determine whether or not a respondent evacuated at any time during the two weeks of the TMI accident.

Loglinear causal modelling, a modified version of path analysis, was employed to determine how well the TMI survey data fit the proposed model of evacuation decision making. In loglinear analysis, the logit formulation, a statistical technique which has been developed for analysing relationships between dichotomous (i.e., two category) dependent variables and either nominal or interval level independent variables, is used to solve a series of structural equations. These make it possible, first, to derive estimates of the direct and indirect effects of variables x_1 to x_{10} on the evacuation decision (x_{11}) and, second, to determine how well the overall model fits the observed data. It should be noted here that because of the requirements of loglinear causal modelling it was necessary to dichotomise all variables before applying the logit technique. A more detailed discussion of loglinear models may be found in Knoke and Burke (1980).

Findings

The final path model of evacuation decision-making based on the TMI survey data is depicted in Figure 12.3. Statistically significant pathways have been shown by solid arrows depicting positive relationships and dashed arrows depicting negative ones. The results of the logit analysis, portrayed in the path diagram, indicate that the evacuation decision during the TMI reactor crisis was influenced by the variables presented in the hypotheses discussed above.

The locational variables of distance and direction directly influenced the decision to evacuate. As distance increased, the rate of evacuation decreased demonstrating an inverse relationship that conforms to classic distance-decay models. It is important to point out, however, that the propensity to evacuate declined only modestly as distance from the site increased out to 12-15 miles, and rapidly only thereafter. Our survey of TMI area residents revealed that there was a uniformly high level of evacuation within 12 miles of the reactor and that the evacuation rate began to

Figure 12.3 FINAL PATH MODEL FOR THE EVACUATION PROCESS DECISION-MAKING AT THREE MILE ISLAND

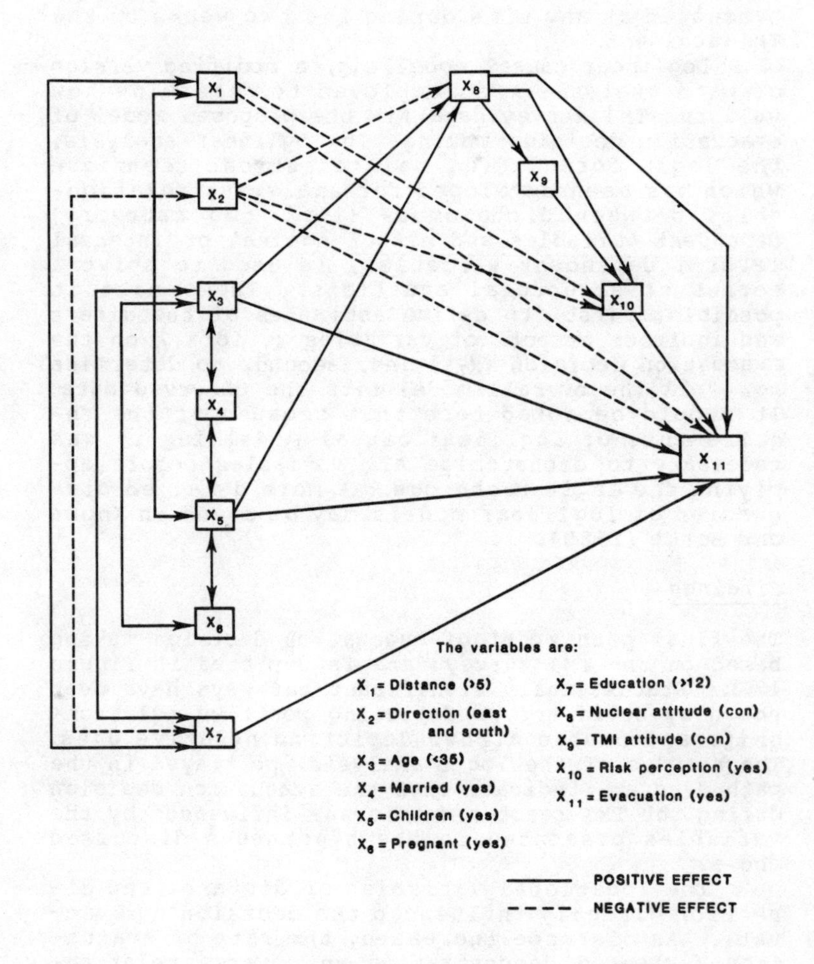

decline noticeably beyond this point (Zeigler, Brunn, and Johnson 1981). Perry (1981) was able to confirm that the perceived threat extended well beyond the five-mile zone, as well. He reported that 50 per cent of the population within five miles of the plant, 50 per cent in the 5-10 mile zone, and 47 per cent in the 10-15 mile zone classified the TMI accident as 'very serious'. In addition to its direct influence, residential proximity also indirectly influenced the evacuation decision, according to our model, through 'TMI accident risk perception'.

Direction from the plant also directly influenced the evacuation decision at TMI. Households south and east of the site were more likely to have evacuated than households north and west. This relationship is depicted as negative only because the direction variable was dichotomised and arbitrarily assigned numerical values. As direction deviated from south and east, the evacuation rate decreased. The more detailed cartographic analysis which we presented in a previous article (Zeigler, Brunn and Johnson, 1981) showed that most of the evacuees avoided the heavily populated regions to the east, which also happened to be downwind from the plant, and opted instead for evacuation quarters in the mountains and sparsely settled communities to the north and west. As depicted in the diagram, direction from the plant also indirectly influenced the evacuation decision through the household head's perception of the risk associated with the TMI accident, and through attitudes to nuclear power.

The NRC data also support our second hypothesis which states that in a nuclear reactor emergency the evacuation decision will be directly influenced by the household's stage in the life-cycle. Two of the four life-cycle indices included in this analysis - age of household head and presence of children in the home - directly influenced the evacuation decision. Families in which the head of household was under age 35 and/or those with young children at the time of the accident were more likely to have evacuated than others. In addition, both of these variables indirectly influenced the evacuation through pre-accident attitudes toward nuclear power. As hypothesised, young families with children were more likely, before the TMI accident, to oppose the use of nuclear power than those over 35 with no children. Neither marital status nor the presence of a pregnant woman directly influenced the evacuation decision; however, both variables exerted an indirect influence through interaction effects with age of household head and the presence of children in the home.

In the NRC survey we also find support for our third hypothesis, that the evacuation decision in a nuclear reactor emergency will be directly influenced by the social status of the household head. From the data set we selected one index of social status, years of school completed. The results of the logit analysis indicate that families in which the household head had completed 12 or more years of school were more likely to have evacuated than those with less formal education. In nearly all of the post-accident behavioural studies this index has emerged as a statistically significant correlate of evacuation. Cutter and Barnes (1982), for example, concluded that 'the more educated the individual, the greater the propensity for that individual to undertake evacuation'. An analysis of TMI data by Hu and Slaysman (1981) also shows that educated households evacuated at a higher rate than their counterparts with less formal education.

Our last two hypotheses were also confirmed by the NRC survey data. The head of the household's pre-accident attitude toward nuclear power and nuclear accident risk perception directly influenced the decision to evacuate. Those household heads in south-central Pennsylvania who felt that the malfunctioning TMI reactor posed an immediate threat to personal and family health and safety were more likely to have evacuated than those who perceived no threat at all.

In sum, the results of the loglinear analysis of the NRC data support all of the hypothesised direct linkages in our model of evacuation decision-making during a nuclear reactor emergency. But exactly how well does the proposed model fit the data? In the loglinear version of path analysis the goodness-of-fit statistic is p^2 (which is analogous to r^2 in ordinary regression). For our evacuation model p^2 exhibits a value of .21 which is indicative of a relatively good fit. (A perfect fit would be represented by a p^2 of one.) In fact, Hensher and Johnson (1981) considered p^2 values between .2 and .4 to be extremely good fits. When the overall model was applied to the individual cases to see how well it could predict the known responses, 71 per cent of the survey respondents were correctly re-classified into the groups to which they were originally assigned on the basis of their actual behaviour during the TMI crisis.

Applying the TMI experience to planning for Sizewell B

The pressurised water reactor (PWR) has been one of the most popular reactor designs in the United States. It was also the type of reactor which came within 280 degrees F. of a meltdown at Three Mile Island. In the United Kingdom all reactors are either MAGNOX or Advanced Gas-cooled Reactors. In 1979, however, only nine months after the accident at TMI, the British Government made public its commitment to the PWR as the basis for a major expansion of the nuclear power industry, beginning with the construction of an additional reactor at the Sizewell site along the coast of Suffolk county, England, where there is already an operating MAGNOX reactor (see Chapter 4 of this volume).

The choice of a technology that was responsible for the accident at TMI was followed by a public inquiry, which began in 1983 and ended in 1985. Safety was one of the issues which assumed centre stage during the proceedings. One of us testified before the Sizewell 'B' Inquiry (Zeigler, 1984) on behalf of the Joint Parish Councils (JPC), an ad hoc committee appointed by the Parish Councils of Middleton-cum-Fordley, Theberton and Eastbridge, and Yoxford, parishes which are among the closest to the site. While the JPC's objections to the Sizewell 'B' unit extended beyond the emergency planning issues raised by the choice of a PWR, one of their chief concerns had been the Government's unwillingness to accept the inadequacy of plans which were already in place for the country's existing reactors. These same plans were considered to be adequate for the PWR scheduled for Sizewell. No changes were deemed necessary since the Central Electricity Generating Board maintained that the British design features added to the Westinghouse PWR would make it safer than any other plant worldwide (Davies, 1984, 24). Needless to say, those closest to the plant viewed such assertions with a sceptical eye and felt that it was important to draw upon the TMI experience for some lessons in the emergency planning arena. Given the public response to the accident at TMI, at least four specific objections may be levelled at the existing plans for an accident at the Sizewell site.

First 'present planning is based on the advice

of the Nuclear Installations Inspectorate that evacuation would not be necessary beyond a radius of 1-1/2 miles from the Power Station' (Suffolk Constabulary, 1984, 5). At the time of the Three Mile Island accident the only plans that existed for the reactor site were for the 2-mile Low Population Zone, and those plans did not include evacuation. Today in the US the evacuation planning zone extends 10 miles from each nuclear installation, and we have argued elsewhere that even this radius is unrealistically restrictive (Johnson, 1984; Zeigler and Johnson, 1984). To plan for an evacuation from only a mile and half's radius from the reactor site is to under-plan for a potential accident for two reasons. First, a release of radioactive isotopes could threaten the population beyond that distance, a fact acknowledged in the British planning documents; and second, people beyond the designated mile-and-a-half boundary are going to evacuate anyway, so it would seem prudent to plan for an evacuation from a much wider area.

Second, according to the plans in existence (Suffolk County Council, 1984, 1) a maximum of 750 people might have to be evacuated in the event of an accident. What this estimate does not take into consideration is the evacuation shadow phenomenon. If we accept the TMI experience as a valid analogy, it is clear that many more people are going to evacuate than will have been advised, and that they will come from a much wider area, a prediction with major implications for traffic management and evacuation time estimates.

Third, according to the Nuclear Installations Inspectorate, potentially affected areas, and therefore the only areas for which evacuation needs to be anticipated, extend 'from a downwind sector up to a distance of 2-3 kilometres from the site' (Nuclear Installations Inspectorate, 1982, 9). Once again the evacuation shadow phenomenon tells us that planning for an evacuation by sectors may not work. People close to the plant in all directions will evacuate even if not advised to do so. If a sector plan is adopted to control an evacuation, there will almost surely be a larger evacuation to control than the plan anticipates.

Fourth, at present all people evacuated from designated sectors will be directed to the Evacuation Centre which will be established at a high school only two miles from the Sizewell site. Is it realistic to believe that the public will be satisfied to put only two miles between themselves and the threatening reactor, when the median evacuation distance at TMI was 100 miles? Public shelters are generally not popular destinations, but people who do use them deserve a safer location situated well beyond the hazard zone.

Conclusions

During the emergency at the Three Mile Island generating station in the United States, evacuation became a common adaptive response among the local population. The spatial behaviours that resulted, however, did not conform to the norms established on the basis of evacuations from natural disasters. Nevertheless, planning for nuclear emergencies in the US has proceeded as if there were no significant differences between nuclear and other types of disasters requiring evacuation. In the United Kingdom, emergency planning for a new generation of pressurised water reactors, about which there has been legitimate safety concern, has been influenced not at all by the experience with the Three Mile Island PWR in 1979. Given British ambitions to expand significantly the number of nuclear reactors in operation and the predicted revival of the nuclear power industry in the US, emergency planning is sure to command the attention of experts in a variety of fields, including geography and the social sciences, over the next several decades. By understanding the geographical dimension of spatial behaviour during the TMI accident we should be able to formulate plans that build on known behaviours in response to a threat that must be conceptualised as demonstrably different from the natural disasters with which we are more familiar.

The magnitude of spontaneous evacuation and the scale of the evacuation shadow phenomenon at Three Mile Island have convinced us that radiological emergency response planning must be built on a base of behavioural information. In this chapter we have drawn upon the survey data compiled for the Nuclear Regulatory Commission in the aftermath of the TMI accident. From it we have selected locational, life-cycle, social, attitudinal, and risk perception variables to explain whether respondents decided to leave or stay behind during the emergency. Using loglinear causal modelling, a modified version of path analysis, we have been able to determine that location was one of the primary variables which influenced the evacuation decision both directly, and indirectly through attitudinal and risk perception variables. In addition, families with young children and those with household heads under 35 years of age evacuated more frequently than their counterparts, and households headed by someone with more than 12 years of schooling were more likely to evacuate than the less educated. These variables were shown in the path analysis to have a direct impact on the decision to evacuate, and two of these variables were also found to exercise an indirect influence by conditioning households' pre-accident attitudes toward nuclear power. Household heads who opposed nuclear power generation in general and the TMI station in particular were more likely to evacuate than their counterparts.

Overall, the path analysis enabled us to determine that the variables which we selected from the NRC data base worked both independently and interactively to influence the decision to leave home in search of safer quarters elsewhere. The model exhibited a relatively good fit, which leads us to conclude that the evacuation response is at least partially predictable on the basis of locational, life-cycle, social status, attitudinal, and risk perception variables that are available for the areas around all nuclear installations or, in the case of attitudes and perceptions, could be collected through social surveys.

The TMI accident has been the US's most serious experience with a nuclear plant accident and therefore is an appropriate analogy for predicting the evacuation response to future nuclear emergencies. In this light, we must accept the need to develop models that will enable us to predict the magnitude of the evacuation shadow phenomenon around other nuclear power sites and estimate its impact on our plans to remove the threatened population from the hazard zone in the minimum amount of time. Rather than depend on education and information control to stifle evacuation response, we believe that evacuation plans need to build on people's natural behavioural inclinations to protect themselves in response to the nuclear hazard.

REFERENCES

AJZEN, I. and FISHBEIN, M. (1980) Understanding Attitudes and Predicting Social Behavior, Engelwood Cliffs, NJ: Prentice-Hall.

AMERICAN PHYSICAL SOCIETY (1985), Radionuclide Release from Severe Accidents at Nuclear Power Plants, New York: American Institute of Physics.

BARNES, K., BROSIUS, J., CUTTER, S. and MITCHELL, J. (1979) Response of Impacted Population to the Three Mile Island Nuclear Reactor Accident: An Initial Assessment, New Brunswick, NJ: Department of Geography, Discussion Paper No. 13, Rutyers University.

BENEDICT, R., BONE, H., LEAVEL, W. and RICE, R. (1980). 'The voters and attitudes toward nuclear power: a comparative study of "nuclear moratorium" initiatives', The Western Political Quarterly, 33, 7-23.

BOWMAN, C. H., and FISHBEIN, M. (1978). 'Understanding public reaction to energy proposals: an application of the Fishbein model', Journal of Applied Psychology, 8, 319-340.

BRUNN, S. D., JOHNSON, J. H., Jnr. and ZEIGLER, D. J. (1979). Final Report on a Social Survey of Three Mile Island Area Residents, East Lansing, MI: Department of Geography, Michigan State University.

CHENAULT, W. W., HILBERT, G. D. and REICHLIN, S. D. (1979). Evacuation Planning in the TMI Accident, Washington DC: Federal Emergency Management Agency.

CUTTER, S. and BARNES, K. (1982). 'Evacuation behavior and Three Mile Island', Disasters, 6, 116-124.

DAVIES, R. (1984). 'The Sizewell B nuclear inquiry: an analysis of public participation in decision-making about nuclear power', Science, Technology, and Human Values, 9, 21-32.

FIREBAUGH, M. W. (1982). 'Public attitudes and information on the nuclear option', Nuclear Safety, 22, 147-156.

FLYNN, C. B. (1979). Three Mile Island Telephone Survey: Preliminary Report on Procedures and Findings, Washington DC: U.S. Nuclear Regulatory Commission.

GOLDHABER, M. K. and LEHMAN, J. E. (1983). 'Crisis evacuation during the Three Mile Island nuclear accident', paper presented at the 1982 annual meeting of the American Public Health Association, Montreal, Quebec (revised).

HANS, J. M., Jnr. and SELL, T. C. (1974). Evacuation Risks - An Evaluation, Las Vegas, Nev.: U.S. Environmental Protection Agency.

HENSHER, D. A. and JOHNSON, L. W. (1981). Applied

Discrete-Choice Modelling, New York: John Wiley and Sons.

HU, T. and SLAYSMAN, K. S. (1981). Health Related Economic Costs of the Three Mile Island Accident, Center for Research on Human Resources, Institute for Policy Research and Evaluation, University Park, PA: The Pennsylvania State University.

INGLEHART, R. (1984). 'The fear of living dangerously: public attitudes toward nuclear power', Public Opinion, 7, 41-44.

JOHNSON, J. H., Jnr. (1984). 'Planning for spontaneous evacuation during a radiological emergency', Nuclear Safety, 25, 186-194.

JOHNSON, J. H., Jnr. (1985a). 'Planning for nuclear power plant accidents: some neglected spatial and behavioural considerations', In Geographical Dimensions of Energy, (ed.), CALZONETTI, F. J. and SOLOMON, B. D., Dordrecht: D. Reidel Publishing Co., 124-154.

JOHNSON, J. H., Jnr. (1985b). 'A model of nuclear reactor emergency evacuation decision-making', Geographical Review, 75, 405-418.

JOHNSON, J. H., Jnr. and ZEIGLER, D. J. (1983). 'Distinguishing human responses to radiological emergencies', Economic Geography, 59, 386-402.

JOHNSON, J. H., Jnr. and ZEIGLER, D. J. (1984). 'A spatial analysis of evacuation intentions at the Shoreham nuclear power station', in Nuclear Power: Assessing and Managing Hazardous Technology, (eds.), PASQUALETTI, M. J. and PIJAWKA, K. D., Boulder, Colo.: Westview Press, 279-301.

KEMENY, J. G., et al., (1979). Report of the President's Commission on the Accident at Three Mile Island, Washington DC.: U.S. Government Printing Office.

KNOKE, D. and BURKE, P. J. (1980). Log-Linear Models, University Paper No. 20, Beverly Hills, CA: Sage Publications.

LIVERMAN, D. and WILSON, J. P. (1981). 'The Mississauga train derailment and evacuation', Canadian Geographer, 25, 365-375.

MEDIA INSTITUTE (1979). Television Evening News Covers Nuclear Energy: A Ten Year Perspective, Washington DC.: The Media Institute.

MEDIA INSTITUTE (1980a). The Public's Right to Know Communicator's Response to the Kemeny Commission's Report, Washington DC.: The Media Institute.

MEDIA INSTITUTE (1980b). Nuclear Phobia - Phobic Thinking about Nuclear Power: A Discussion with Robert L. Dupoint, M. D., Washington DC: The Media Institute.

MELBER, B. D., NEALEY, S. M., HAMMERSLA, J. and RANKIN, W. L. (1977). Nuclear Power and the Public: Analysis of Collected Survey Research, Seattle, WA: Battelle Human Affairs Research Centers.

NRC/FEMA (1980). Criteria for Preparation and Evaluation of Radiological Emergency Response Plans and Preparedness in Support of Nuclear Power Plants, Washington, DC.: U.S. Nuclear Regulatory Commission and Federal Emergency Management Agency (NUREG-0654).

NUCLEAR INSTALLATIONS INSPECTORATE (1982). Emergency Plans for Civil Nuclear Installations (NII/5/54).

OTWAY, H. J. (1977). 'Risk assessment and the social response to nuclear power', Journal of the British Nuclear Energy Society, 16, 327-333.

OTWAY, H. J. and FISHBEIN, M. (1979). 'The determinants of attitude formation: an application to nuclear power', International Institute for Applied Systems Analysis, Research Memorandum, RM-76-80, Laxenburg, Austria.

PERRY, R. W. (1981). Citizen Evacuation in Response to Nuclear and Non-Nuclear Threats, Seattle, Wash.: Battelle Human Affairs Research Centers.

PERRY, R. W. LINDELL, M. K. and GREEN, M. R. (1981). Evacuation Planning in Emergency Management, Lexington, MA.: Lexington Books.

PLOSILA, H. W. (1979). The Socio-Economic Impacts of the Three Mile Island Accident, Harrisburg, PA.: Governor's Office of Policy and Planning.

ROGOVIN, M. and FRAMPTON, G. T. (1980). Three Mile Island: A Report to the Commissioners and to the Public, Nuclear Regulatory Commission Special Inquiry Group, Washington, DC.: Nuclear Regulatory Commission.

SLOVIC, P., FISCHOFF, B. and LICHTENSTEIN, S. (1979). 'Rating the risks', Environment, 21, 14-39.

SLOVIC, P., FISCHOFF, B. and LICHTENSTEIN, S. (1980). 'Facts and fears: understanding perceived risks', in Societal Risk Assessment: How Safe is Safe Enough? (eds.), SCHWING, R.C. and ALBERS, W. A. New York: Plenum Press, 181-214.

SUFFOLK CONSTABULARY (1984). Sizewell Power Station-Police Emergency Plan (LPA/S/92).
SUFFOLK COUNTY COUNCIL (1984). Sizewell Nuclear Power Station Emergencies (Local Authorities Involvement).
WEBBER, D. J. (1982). 'Is nuclear power just another environmental issue? An analysis of California's voters', Environment and Behavior, 14, 72-83.
ZEIGLER, D. J. (1984). 'Evidence in Chief', Sizewell B Inquiry Before Sir Frank Layfield, Day 296, Snape, Suffolk County, England.
ZEIGLER, D. J., BRUNN, S. D., JOHNSON, J. H., Jnr. (1981). 'Evacuation from a nuclear technological disaster', Geographical Review, 71, 1-16.
ZEIGLER, D. J. and JOHNSON, J. H., Jnr. (1984). 'Evacuation behavior in response to nuclear power plant accidents', Professional Geographer, 36, 207-215.
ZEIGLER, D. J., JOHNSON, J. H., Jnr. and BRUNN, S. D., (1983). Technological Hazards, Resource Publications in Geography, Washington, DC.: Association of American Geographers.

PART 4:

FUTURE DIRECTIONS

Chapter Thirteen

PROSPECTS FOR THE NUCLEAR DEBATE IN THE UK

Timothy O'Riordan

Overview

On Saturday 26th April an event took place that was to have a major influence on the global nuclear debate. That event was the near meltdown of the Russian RMBK graphite moderated boiling water pressure tube reactor No. 4 at Chernobyl in the Ukraine. As the world now knows, the reactor exploded, spewing radionuclides in a fire-driven cloud high into the atmosphere. Unpreparedness, coupled with politically-inspired poor communications between the Ukraine and Moscow, and an almost total inability to know what to do to control the emissions and to evacuate the large neighbouring population, led to confusion and criminal negligence in under-reporting the accident and denying those most at risk the information and resources to save themselves. The result was a radiation plume, fuelled by a fire that burned almost uncontrolled for nearly five days, which spread literally throughout the northern hemisphere. Radiation fallout created near panic amongst an anxious and scientifically ill-informed European people, causing some governments to ban the sale of milk and vegetable products, especially from Eastern Europe, but also from within their own borders. In parts of Britain milk consumption dropped by 20 per cent as anxious parents denied their children their daily ration for up to three weeks. In the Netherlands dairy farmers were forbidden by law to allow their cattle to graze on the nutritious spring growth, being forced instead to buy in expensive food concentrates. The Russians refused to pay compensation on the grounds that these actions were based on panic and ignorance, not on scientifically proven danger.

Table 13.1

ATTITUDES TO FURTHER DEVELOPMENT OF NUCLEAR POWER IN THE 10 EEC MEMBER STATES[a]

Percentage of respondents indicating worth of further development.

	Worthwhile			Unacceptable risk		
	1978	1982	Difference	1978	1982	Difference
Belgium	29	27	-2	39	37	-2
Denmark	37	25	-12	34	49	+15
France	40	51	+11	42	31	-11
Federal Republic of Germany	35	37	+2	45	30	-15
Greece	-	15	-	-	50	-
Ireland	43	13	-32	35	47	+12
Italy	53	34	-19	29	43	+14
Luxembourg	35	32	-3	31	49	+18
The Netherlands	28	34	+6	54	48	-6
United Kingdom	57	39	-18	25	37	+12

a

Total sample size 9700 (1982). Greece was not included in the 1978 survey. Respondents were presented with four response categories to indicate their attitude concerning the development of nuclear power stations. These were: 'it is worthwhile', 'no particular interest', 'the risks involved are unacceptable' and 'don't know'. This table is based on findings reported by the Commission of the European Communities (1982).

Source: van de Pligt, 1985, p.89.

Chernobyl is one more household name that symbolises environmental disaster and technological failure. It stands on a par with Torrey Canyon, Seveso, Three Mile Island and Bhopal. For the nuclear industry one can confidently speak of a 'post Chernobyl era', for Chernobyl was much more than a nuclear nightmare. Chernobyl has irreversibly shifted public opinion against nuclear power generation. Chernobyl has eroded the trust that many used to place in accident scenario predictions and fail-safe technology. Chernobyl has weakened the public faith in scientific prowess and undermined the role of expertise. Chernobyl has made it impossible for the nuclear industry to be its own advocate. Perhaps most significantly, Chernobyl has caused the public to believe anything about the dangers of nuclear power, and to be almost immune to arguments that the industry is safer than any other form of electricity generation, or other dangers faced by ordinary people in everyday life. Chernobyl also demonstrates that the global nuclear club is branded collectively by the mistakes of its most negligent member. The industry worldwide has felt the repercussion of an event that, until 26 April 1986, 'could not happen here'.

But Chernobyl must be put in perspective. Nuclear power and radioactivity have long been topics which raise strong passions. Over the past dozen years the balance of public opinion has swung against nuclear power generally and radioactive waste management in particular. Table 13.1 shows that for the 10 member states of the European Community this trend is not universal, though it is most prominent in Denmark, Iceland, Italy, Luxembourg and the UK. After Chernobyl, polls throughout Europe revealed a two to one opposition to further expansion in nuclear power, though the French remain stoically in favour of the technology. The UK Social Attitudes Survey of 1985 (Central Office of Information 1984) suggest that, of various pollution sources, radioactive waste is seen by the public as having the most serious effect on the environment (Table 13.2). Studies from elsewhere, summarised by Kasperson (1984) show a similarly high level of public antagonism to radioactive waste disposal.

This shift has been swept along by powerful undercurrents. The rise of the peace and green movements, and the growth of more radical environmentalism, flourish in the antagomism felt over state commitments to military and civil nuclear

Table 13.2 WHAT EFFECT ON THE ENVIRONMENT DOES IT HAVE? (All figures in %: Don't Knows omitted)

	Very Serious	Quite Serious	Not Very Serious	Not at all Serious
Nuclear power waste	69	18	9	2
Industrial effluents	67	25	6	1
Industrial air pollution	46	40	11	2
Lead from petrol	45	39	11	2
Traffic noise/ dirt	20	45	29	4
Aircraft noise	7	24	50	17

Source: Central Office of Information (1985)

technology. A number of mainstream political parties have turned or are turning anti-nuclear, opposed both to the military and civil arms of the industry. In the UK, the Labour Party is having difficulty in defining its nuclear platform, because some powerful trade unions have long supported the industry, but in this post-Chernobyl era, nuclear politics have been catapulted into the electoral arena. The main opposition parties will have to establish a consistent and credible stance on nuclear power and radioactive waste disposal. Western society has become a fragmented nuclear culture in which opinion is deeply divided, and prejudice and misunderstanding are ripe for exploitation.

The aim of this concluding review is to summarise the main themes of the preceding chapters by setting them in the context of the dynamics of contemporary nuclear politics. A particular feature of this book has been the examination of the relationship between the state, the nuclear industry and the anti-nuclear lobby. The chapters have all been written against a background in which the state has supported, even encouraged, the nuclear industry with generous financial and research assistance. Despite Chernobyl, there is as yet no sign at present of this relationship being altered. If anything, the opposite is the case. States are beginning to collaborate internationally on research and demostration projects for the next generation of nuclear reactors. This involves the fast (breeder) reactor which converts enriched uranium into plutonium; and nuclear fusion, where energy is created by bundling deuterium and tritium molecules into hot, plasma-like gases to create enormous amounts of energy some 10 million times more concentrated than similar reactions involving hydrocarbon fuels. The fusion reactor is the ultimate goal of the nuclear lobby as it is supposed to provide almost limitless amounts of power. But, as with the fast reactor, the technology is by no means assured. The research and development costs are astronomic and can be secured only by the combined resources of several states acting together.

The four ages of nuclear power

One can discern four ages in the life of civil nuclear power in the UK over the past thirty years. These can be characterised as the age of innocent expectation (1946-1966), the age of doubt (1966-1974), the age of anguish (1974-1981) and the age of public justification (1981 to the present). The first period was associated with the post-war exuberance over nuclear technology. The British led the civil reactor field, and were able to enjoy the apparent luxury but economic burden of designing and constructing individual reactors - the first generation MAGNOX reactors. The second period began with the decision to construct the advanced gas cooled reactor (AGR), the successor to the MAGNOX. But the AGRs were inadequately designed and suffered many delays and cost over-runs. The AGR experience began to raise doubts about the cost effectiveness of nuclear power, and about the accountability of the nuclear industry to public opinion and democratic parliamentary procedures. This point is developed by Fernie and Openshaw in Chapter 5.

During the third age, that of anguish, the screws began to tighten. The industry was not granted the pressurised water reactor (PWR) that many cherished, though support for the steam generated heavy water reactor (SGHWR) soon evaporated. The safety and cost of the PWR became key issues. A number of informed commentators and analysts threw doubt on the integrity of the safety record and probabilistic risk assessment of fault sequences and accident scenarios. Furthermore, in the aftermath of the oil price rise and industrial recession, energy forecasts made in the heady days of the early 1970s proved massively over-optimistic throwing further doubt on the claims for large programmes of reactors. Finally the Three Mile Island incident shocked the industry, the safety regulators and public opinion. All this resulted in a completely new appraisal of the safety and economics of civil nuclear power. The confidence that characterised the men behind the industry, began to sap.

The fourth age is both fascinating and frustrating. Nowadays the industry recognises that it has lost public confidence. Senior managers are genuinely mystified by this and turn to patronising arguments of scientific rationality and conventional public relations gimmicks to overcome the credibility gap. These tactics are counter-productive. The more the industry tries to sell its case, the less the public accepts its propaganda. The battle that is raging nowadays is a struggle for trust and credibility. Chernobyl has tipped the scales, for the time being at least, against the industry's fortunes. Any further major events, always a possibility, will load the scales even further. In this context, the media play an extraordinarily important role. They thrive on disaster and novelty yet, for the most part, journalists are scientifically illiterate. Cautious experts, who can put nuclear matters into perspective, decline to be interviewed unless sure of their facts. Less cautious commentators, unencumbered by all the relevant details, are tempted to pronounce and they achieve much publicity. Populist science feeds media science and dispassionate analysis is the loser.

Six themes

The contributors to this book address six themes:

1. Nuclear decision making and democratic accountability

In order for the nuclear industry to progress,

development at each stage of the nuclear fuel cycle must be maintained. The result is a kind of interactive momentum that can both accelerate or stall nuclear evolution. These stages are uranium mining, milling and fuel fabrication, reactor operation, waste fuel reprocessing, radioactive waste management and spent reactor decommissioning. No stage can be allowed to lag behind as the industry strides forward. Popular protest and political opposition has to challenge both the whole cycle and every part. About half of the 40-50 per cent of the public who form the anti-nuclear camp in the UK (O'Riordan, 1985) are concerned that what they regard as democratic political processes may not be fully responsive to such protest. Democratic procedures seem neither sufficiently strong willed nor independent to stop the seeming inevitability of nuclear expansion. This line of argument appears in Chapters 2 and 3 in terms of energy policy formulation, and in Chapters 4, 6, and 7 in terms of energy management institutions. All their authors suggest that sustained public opposition to nuclear developments can at least modify policy and shift political attitudes. It remains to be seen, however, whether nuclear developments can actually be stopped by protest. Nuclear research, development and demonstration may be stretched, delayed, even mothballed, but so far not curtailed.

2. Nuclear hegemony and a narrow-based energy policy

Financial commitments to nuclear power are so vast that alternative means of generating electricity or of reducing electricity use by improving efficiency of use, or of simply managing the electricity network more sensibly, are not being properly considered. In the UK this is a particularly controversial policy issue. The House of Commons Energy Committee (1984) reported that in 1984-85 only £45.5 million was spent directly on non-nuclear energy research and development (renewable fuels, synthetic fuels, combined heat and power) compared to £196.6 million commited to nuclear energy research and development (R & D). The Committee criticised the Department of Energy for sanctioning a huge imbalance between R & D in energy supply technologies (£235 million in 1984-85) compared to investments in the technologies and practices of energy conservation. R & D in the so-called demand reduction aspects of energy amounted to only £10 million in 1984-85. In short, the nuclear bias loads national energy policy so that realistic alternatives are not permitted to flourish. Accordingly, they do not appear technologically credible or economically viable. This point is raised in Chapters 3 and 4, which conclude that distortions in policy appraisal are so embedded in the energy policy making system as to be almost unobservable even to those responsible for them.

3. Scientific discredit and popular expertise

Scientific valuations of nuclear related dangers no longer command unquestioning public acquiescience. This point is particularly stressed in Chapters 6 and 7. They show that, in many cases, scientific analyses are simply not able to produce unambiguous results. There are too many unknowns, the data are insufficient or unavailable and the methodological tools are found wanting. The science of prediction is not really a science at all: it is an art form of educated guesswork. Chernobyl has exposed its failings; failings that normally remain hidden, but which are never absent. The upshot is that expertise does not carry as much weight in public and political judgement as it once did.

An important reason for this doubting of expertise is the rise of non-establishment science. This is challenging the very heart of conventional science - its assumptions, its theories, its models and its predictions. An interesting offshoot of the nuclear industry has been the elopement of some of its prominent scientists into the anti-nuclear camp. They give the anti-nuclear lobby well-based information, measured judgement and a degree of credibility that is more observable in this arena than in other aspects of environmental politics. The result is often an outburst of slanging between supposed experts that mystifies and alienates the outsider. Alternatively, expertise of whatever kind simply does not sway the intuitive judgements and feelings of lay people. As already mentioned, the contemporary nuclear battle is the fight for public confidence and credibility, not scientific or technological justification. This is slowly dawning on the nuclear industry, which is still uncertain as to how to handle the problem.

4. Fusing science with social values

The need to bring social values into scientific appraisals of nuclear technology places special burdens on assessment methodologies and institutions. This issue is addressed particularly in Chapters 4 and 5 though it is also pertinent to the two chapters on radioactive waste management, Chapters 7 and 8. The longer term question of how to deal with a decommissioned nuclear reactor, discussed in Chapter 9, is also relevant.

Two problems surface here. One is how best to undertake economic assessments of nuclear-related capital projects. The time pattern of costs and benefits is heavily asymmetric in favour of benefits, especially the early benefits. In addition, quantifying in money values such costs as radiation contamination from normal operations or from accidents, is almost impossible to undertake faithfully. So economic assessments of nuclear investments tend to favour the immediate and tangible benefits of more electricity now, rather than the far less predictable costs which may span out over many generations. This raises the difficult issue of discounting the cost and benefit streams of nuclear investments. Discounting is a procedure for equating gains and losses to a common point of reference, normally present value money. But when costs may emerge in the distant but unpredictable future, discounting can become meaningless. Since there are no adequate methodologies for dealing with such non-equivalent phenomena, this places assessment institutions such as public inquiries or special commissions of investigation in something of a dilemma. That dilemma is how to marry quantifiable criteria prepared through conventional cost benefit analysis, with unquantifiable elements that do not conform to known methodologies. This particular problem plagued the Sizewell B Inquiry (Chapter 4) though it is not confined to cost-benefit analyses of nuclear investments.

The other difficulty is the weighing of acceptable risk, an elusive concept with no real meaning (O'Riordan, *et al.*, 1985). It is a function of ethical and distributional considerations of who is exposed to dangers, when and how seriously. But it is also an outcome of effective participation by ordinary people in any final judgement about the dangers being faced. In short, risk becomes acceptable only when its context is fully understood, alternatives to the risk creating source are roundly evaluated, and the resulting social benefits command public approval. As the Sizewell B Public Inquiry discovered, this is an extraordinarily difficult task. Yet if public confidence in nuclear power has to be won, this has to be done. For all its failings, outlined in Chapter 4, the Sizewell B Inquiry is to be commended for tackling, if not overcoming, this challenge.

5. Managerial competence

Lay people will never know enough about curies, or sieverts, or probabilities, or fracture mechanics to be able to understand a technical risk assessment, whether that assessment be of a nuclear reactor, an engineered waste disposal facility or the decommissioning of a retired nuclear plant. Chernobyl served to prove that the public are mystified and frightened of radiation because they do not understand it. Media images of radiation clouds drifting menacingly across the North Sea aggravate these fears, in the face of which scientific reassurance is largely ineffective.

What people want to know is how reliable is the management of these processes. By management, one is referring to the safety licensing process, component fabrication checks, construction standards, operator performance and regulatory surveillance. In the UK most of these management roles are performed behind closed doors by people upon whose competence and commitment the public and the industry have implicitly to rely. However, no longer are the public, as bystanders, willing to trust nuclear facility management. A 1982 poll conducted for the European Community discovered that 37 per cent of those questioned expect nuclear accidents to occur from a combination of human error and technical failure, while 27 per cent believe that human error alone is the most likely cause (Sizewell B Inquiry Transcripts, Day 301, p. 47 at B to C). Post Chernobyl, where human error compounded technological failure, these beliefs have shifted towards further loss of faith in managerial competence.

Fernie and Openshaw, Blowers and Lowry and Armstrong suggest that all is not well in the world of nuclear regulation. Standards are not always adhered to, the regulators are by no means aware of what is going on, and the public are almost totally ignorant of the whole management process. Furthermore, Fernie and Openshaw point out that the traditional mechanisms through which nuclear regulation has in the past been conducted in the UK, namely administrative discretion, generalised principles relying on judgement and experience of their interpretation, and secrecy, are all being challenged. The call for greater openness and formality in nuclear regulation is almost universal.

Chapter 12 points to one particular failing. This is the assumption, built into emergency and evacuation plans, that people will wait until they are told what to do when a nuclear emergency is called. Zeigler and Johnson argue that this is wishful thinking on the part of the managers. People will not behave as predicted, but may well act recklessly, in panic and without control. If nuclear management is to command public confidence it must demonstrate that all plans and operational procedures take into account random human error and unpredictable behaviour patterns. This is a daunting but necessary task. It is worthy of note that the Inspector at the Sizewell B Inquiry asked the Nuclear Installations Inspectorate, the UK Government nuclear regulatory agency, to prepare an accident scenario for a 8km zone. The emergency plan normally made is for only a 3km zone. The Inspector wanted reassurance that emergency services could cope over a much larger area and with a bigger population. His motive was to assuage local anxiety that people could not be safeguarded in the event of an extraordinary accident.

Post Chernobyl, even this is not enough. Chernobyl showed that in the event of a large disaster the population of an area of up to 15km radius would need to be evacuated, but that over a radius of possibly 50km people would spontaneously seek to flee. The implications for emergency evacuation are almost unimaginable. Within 15km of the Hartlepool AGR, for example, live nearly half a million people. Over a million reside within 50km radius of the plant (Openshaw, 1986, pp. 261, 282). It is highly likely that emergency procedures will now have to be rewritten and re-enacted.

6. The geography of nuclear power

Nuclear facilities have to be located somewhere. Wherever that is, part of the population is exposed to a different quantum of peril than the rest. Some people (about 40 per cent in Britain) do not mind this: they are convinced that nuclear power is a positive social good. Others (about 20 per cent) will learn to live with nuclear facilities if they do not go wrong, bring in jobs and economic activity, and are operated by friendly people who become part of the local community (O'Riordan, 1985). For some, at least, familiarity and economic dependency breed acceptance and support. MacGill and Phipps (Chapter 10) have found evidence of this in their study of the attitudes of West Cumbrian people to the BNFL complex at Sellafield. Even when the plant went wrong and discharged radioactive materials into the sea to be subsequently washed onto the local beach, few, whose livelihood was connected to the plant, were critical. The nuclear industry is anxious to breach the barrier of unfamiliarity and fear, for it believes that once this is crossed, and so long as the industry manages its affairs well, the people will come to support it.

But the MacGill-Phipps data show that those who do not earn a living from BNFL are far less likely to favour the complex. Furthermore the chapters on radioactive waste management, notably Chapters 6 and 8, reveal that an unfamiliar technology in a greenfield site which does not bring with it demonstrable economic gain and whose presence may drive away potential employers - such a technology is guaranteed to generate wholesale and cross-party opposition. The alien technology becomes a symbol for blame and frustration. Lee and Brown (1985) show that two fifths of a sample of the population in areas likely to be selected for radioactive waste disposal sites ranked health risk as their main concern. They wanted to see both health and other kinds of risk fully explored in any public inquiry into the proposal (Table 13.3).

The geography of nuclear power is the geography of familiarity and the geography of dependence. It is tempting to conclude that once one major nuclear facility is in place it could well be the forerunner of a nuclear complex. Both American and British utilities are pursuing a policy of constructing reactor clusters; possibly to the point of designing new reactors adjacent to or around the hulks of decommissioned 'grandparents'. So the geography of nuclear power may also prove to be the geography of nuclear permanence.

This in turn raises an interesting question. About half of those who tend to favour nuclear power generally, oppose nuclear developments in

Table 13.3

OBJECTIONS TO THE SITING OF RADIOACTIVE WASTE (%)

health risk	40
leaks-spills	12
unsafe	7
too near towns	6
fear of unknown	4
petitions would be raised	12

ISSUES TO BE RAISED AT A PUBLIC INQUIRY
(n = 1511)

		Score
	health effects	3540
frequencies weighted	unforeseen risk	1374
x3 for first choice	effect on ecosystems	906
x2 for second choice	suitable geology	889
x1 for third choice	local opinion	709
	transport distances	531
	national interest	249

Source: Lee & Brown, 1985, 538

their locality. The majority of this group are more concerned about the disruption caused by construction (noise, dirt, congestion and a temporary labour force) than by the facility itself (van de Pligt, 1985). If nuclear reactors are to be clustered, their construction nuisance could almost be permanent, certainly for the lifetime of middle aged residents. This could be a key factor that opponents to major nuclear developments might learn successfully to exploit, because it might widen the basis for opposition.

Future tactics by the nuclear industry

All these themes emphasise the point made at the outset. The contemporary nuclear debate is about gaining legitimacy, through authoritative support for the technology, for its economics, for its safety, for its managerial reliability and for its siting. Legitimacy can be won by fair means or foul. The discreditable way is to maintain a tight political stranglehold, to lock governments and the financial institutions into inextricable commitments, and to operate behind a cloak of secrecy and technical mystification. For approximately 15 years the nuclear industry successfully followed such a strategy. In so doing, it became big in size, big in ego, excessively self-assured and paternalistic towards doubters. These days are now gone, though there are many in the industry who still yearn for the past.

Today, the industry realises that it has to make its case. It has to open up its technology, its safety assessments, its economic justification and its siting strategies. It also has to listen. It has to respond to protest, to adverse and often distorting analysis in the media, and to the demands for safety standards and managerial excellence that are all but impossible to meet. The problem is that the more it exposes itself and the more it responds to accusation, the more it finds itself embroiled in argument and disillusionment. Ironically, the industry is now prey to its legacy of secretiveness and apparent complacency.

As matters stand, the political survival of the British nuclear industry to some degree depends upon the retention in power of the only major political party, the Conservatives, that unequivocally supports it. Even after Chernobyl the Tories were able to proclaim:

> The Government believes firmly that nuclear power, subject as it is to the most stringent safeguards, has an essential contribution to make to the provision of electric power, economically and at a level of risk comparable to or better than that posted by alternative sources of power (Department of the Environment, 1986, p. 2).

In a Commons debate on 13 May 1986 (Official Report, 97, No. 112) the Labour Spokesman, Dr. Cunningham, proclaimed that the Labour Party envisaged 'no case for ordering any nuclear power stations in the prevailing circumstances' (col. 585). This was hardly an outright rejection, and has caused much controversy within the Party. Subsequently the Labour Party leadership has commited itself to a non-nuclear energy policy. The Liberals were more specific. Their spokesman, Malcolm Bruce stated:

> The Liberal Party believes that investment in energy efficiency, improvements in the environmental acceptability and the efficiency of

> coal and oil, coupled with the development of alternatives, could meet all our energy needs, and would allow for a gradual changeover to a non-nuclear energy mix. This is attainable in the United Kingdom (col. 593).

However, it remains to be seen whether any party with the responsibility of office is prepared to kill the industry outright. There is more rhetoric than reality about party manifestos and nuclear power.

How might the nuclear industry respond? In Britain in particular it is going to rely a lot, in the first instance, on the outcome of the Sizewell B Inquiry, on the Dounreay Inquiry into the European Demonstration Fast (Breeder) Reactor Reprocessing Plant (a five-country collaboration project if it is commissioned), and on the public inquiry into a suitable site for the disposal of low level radioactive waste (as discussed in Chapter 6). Each of these inquiries will be very different. Each will be highly politicised as the conflict between proponents and opponents intensifies.

The role of the Parliamentary Select Committee should be especially interesting to observe, as the Chairmen of both the Energy and Environment Committees are both anxious for their members to be part of the final decision making process on both the Sizewell B and the low level waste issues. For the foreseeable future, Parliament is likely to be involved in nuclear choices to a far greater extent than ever before, though, paradoxically, this political participation may take place at the expense of local consultative procedures.

One possible future development is to take the oversight of the industry out of its own hands. It may now be necessary to establish some sort of totally independent nuclear oversight commission, composed of respected technical and lay people, reporting annually to Parliament on all matters nuclear. Such matters might include safety standards, regulatory guidelines, radiobiological developments, siting criteria, accidents and near accident reports. This commission would be quite a different body from the existing advisory committees such as the National Radiological Protection Board, the Advisory Committee on the Safety of Nuclear Installations and the Radioactive Waste Management Advisory Committee. All these bodies would continue to perform their existing functions, but the oversight commission would carry out a different kind of task - overseeing and reporting on the work of these bodies and examining matters of public interest. The Committee's purpose would be to act as an authoritative and credible guardian of the public interest vis à vis the nuclear industry. Such a dramatic development might have to be necessary if the industry is to win the battle for legitimacy.

In all probability the nuclear industry will reject such an idea, at least for the time being. It wants to put its own house in order and be responsible for its own affairs. So it is more likely that it will pursue a number of tactics, all associated with the public image, and all of which could misfire.

First, the nuclear industry may step up its public relations and public information activities, probably by consolidating the united front of the whole industry. The danger here is that by collaborating across a broad front the industry may become damned by its own association. The more cohesive the industry appears - the more mutually supporting - the more the scope for collective downfall if things go wrong and the public credibility battle is lost.

Second, the industry might spend a lot of money trying to 'correct' public 'misapprehensions' and in 'clarifying' 'misunderstandings'. This would involve exhibitions and leaflets aimed at educating people about nuclear technology and nuclear physics, about comparable risks in other electricity generating plants, and elsewhere in daily life, and about the 'cleanness' of nuclear power generation compared to coal and renewables. Such a tactic could lay the industry wide open to countervailing scientific challenge.

Third, the industry may seek to provide compensation for communities where future reactors and waste disposal sites may be planned. That compensation could be in the form of community facilities such as new by-passes or golf courses, or it may emerge through offers to buy property blighted by protest and anxiety.

It is even possible, though unlikely, that the Government may change its policy toward local authority rates and enable local communities to get all or part of the equivalent of the rateable value of the nuclear facilities. Presently this money is sent straight to the Exchequer, and the host communities receive nothing. In the case of the Sizewell B PWR, if built, this amounts to some £3 million annually. The French are less shamefaced about compensation, as Chapter 3 indicates. In the past Electricité de France has reduced the tariff for electricity consumers living near nuclear stations. But this practice has subsequently been declared illegal, so the French are considering lump sum payments to host communities instead. The Government has responded positively to a recommendation by the House of Commons Environment Committee (1986, p. cvii) in its report on radioactive waste, that it should examine seriously the whole issue of compensation. So this is now firmly on the political agenda.

Fourthly, the industry may seek to open up its safety case and management practices to more independent scrutiny. This might partly be done within the public inquiry process. However, the decision by the Scottish Secretary not to allow the safety case for the proposed Dounreay plant to be heard is not encouraging in this respect. Ostensibly that decision was taken because only outline planning permission is required. It is likely that this decision will be widely misinterpreted and the inquiry will become a focus for dissent. Greater openness might also be achieved by exposing the whole regulatory process to review and regular reporting. It is unlikely that the industry would wish for this, but it may have to accede to it.

Finally, and more controversially, the industry may seek to gain legitimacy by showing that its product is genuinely needed. To achieve this involves an interesting paradox. The electricity supply industry would have to invest in more research and development into improved electricity load management and electricity use efficiency (see Chapter 2) - as advocated by the Electricity Consumers Council and the Council for the Protection of Rural England at the Sizewell B Inquiry. Such an approach would result in lower demand forecasts and a greater delay in plant ordering. But it should demonstrate that there is still a shortage of base-load power that is genuinely needed. Then the matter really falls on the politics and economics of more nuclear versus more coal. This is also a high risk strategy but may prove necessary if the complexion of Parliament changes following the next general election.

Future tactics for the anti-nuclear lobby

Much of what has just been written is speculation arising from points raised in this book. The anti-nuclear lobby should be able to counter many of these possible developments with varying success.

The primary tactic for the lobby is to continue to generate political pressure against further development of nuclear power. Chernobyl has played into its hands at a critical time of political choice. The Sizewell B decision is now an electoral issue: delay would probably smother the project since inaction coupled with political uncertainty would make it extremely difficult to maintain a viable project organisation. The suspended animation already imposed on the construction teams in the long twilight of the Inspector's write-up (lasting 18 months) is costing the CEGB £3 million per month.

High profile political lobbying by the anti-nuclear campaigners would pay dividends if a future Parliament had no clear Conservative majority. That lobbying would require painstaking critical argument, not emotion-ridden prejudice. The arguments that might be most telling would be those for improved electricity management, investments in energy conservation and, especially, electrical use efficiency; for further work on the link between economic restructuring and energy demand; for increased attention to combined heat and power; and for technological improvements to lengthen the life of existing generating stations.

Sooner or later, however, new supply sources will have to be found. By then the anti-nuclear lobby will have addressed the future of the coal industry. This will prove to be one of its most challenging tasks, because the economics of coal pose some very awkward issues indeed. These include the efficiency of the technology and the costs of sulphur dioxide and nitrogen oxide removal from existing coal-fired stations, the higher costs but long term technological advantages of fluidised bed combustion, the technical and geological problems of ensuring future coal supply, and the social consequences of restructuring the coal mining industry into the 21st century. These are not topics that the anti-nuclear lobby can duck, but they are issues where some better analyses and more imaginative thinking are required. By raising and widening the quality of that debate, anti-nuclear interests would provide an important service for the nation. While there is certainly a future role for renewable energy sources, this may always be a modest one. Nonetheless, the lobby will have to show how any significant development of renewable energy resources is neither environmentally damaging nor cost ineffective. A Severn Barrage is now an option that is being actively pursued, as is photo-electric cell technology.

In opposing the back end of the radioactive fuel cycle, the anti-nuclear lobby has an equally challenging and quite different task ahead. On waste fuel management, it must demonstrate that reprocessing should cease, and hence why much of what is happening and is proposed at the Sellafield complex is neither economically viable nor necessary. The House of Commons Environment Committee (1986, p. lxxxviii), heard that the net cost reprocessing spent AGR fuel was around £300,000 per tonne - over 50 per cent higher than the estimate of the Windscale Inquiry Inspector, and way beyond the currently falling price of enriched uranium. That Committee also heard evidence that indefinite storage of spent fuel was both technically possible and socially more desirable. Storage facilities could be extended where waste disposal and storage arrangements already exist, namely Drigg in Cumbria for low level waste, and Sellafield for more active wastes. By eliminating the reprocessing requirement, irradiated spent fuel transport would virtually end, and the quantity of the more radioactive wastes would fall dramatically. Nuclear reactor sites could become total complexes of nuclear management-generation, waste storage and decommissioned plants, and relatively few sites would be required. The task of containing the future of nuclear energy is manageable though a lot of attention would have to be given to environmental appraisal, community compensation and long term economic encouragement to local economic enterprise. With respect to Sellafield, Greenpeace has begun to address this critically important issue (Nectoux, _et al_., 1985).

On decommissioning, the anti-nuclear lobby should demonstrate even more clearly how undeveloped is the technology and scientific understanding of the problem. This point is touched on in Chapter 9. It should force the industry to undertake much more research on what materials can safely be removed from disused reactors and what cannot. The final decommissioned hulk should also be a target for much more imaginative investment in architectural design and landscaping.

Probably the strength of the anti-nuclear lobby lies in the postulated but credible association between civil and military use of nuclear power, the aggravatingly high wall of secrecy that surrounds the whole issue of plutonium inventories and reprocessing, and the danger of proliferation. In this arena secrecy is proving to be a vote losing strategy. The lobby can and should exploit this mercilessly, for it is a theme that causes much misgiving amongst the uncommited middle ground.

Finally, the anti-nuclear lobby will have to address the issue of future commitments to international consortia in fast reactor technology and fusion power. In many respects this may prove to be its greatest challenge because the state is so deeply involved in such developments. International technical and political obligations are very powerful. The problem will be to distinguish between research for its own sake, and research that will lead to a demonstration and possible commercial development. They may be impossible to separate in practical terms. Thus the ultimate legacy of the anti-nuclear lobby must be the curtailment of all nuclear and related research. The wheel will have turned full circle from the optimistic post-war era when peaceful uses of nuclear power were regarded as the salvation of civilisation. The anti-nuclear lobby will have to show that civilisation can flourish only in a non-nuclear world. In the battle for credibility and legitimacy, the stakes are indeed very high.

In the final analysis, however, all of these tactics by the anti-nuclear lobby could assist the credibility of the nuclear industry and ensure its survival. The constant critique is helping to sharpen up safety standards and regulatory requirements, the overseeing bodies are becoming more inquisitive, and the economics of comparative energy utilisation are beginning to even out as common standards of environmental protection and public safety are sought. If the nuclear industry responds constructively and creatively it may forge a new liason both with the state and the public to produce a slimmed down nuclear programme, coupled with more imaginative electricity management measures, with innovations in community relations and with the functional separation of civil and military nuclear streams. These, however, are only possibilities. Also possible is that the nuclear industry will fall victim to its own failures.

REFERENCES

CENTRAL OFFICE OF INFORMATION (1984) Social Trends, London: HMSO.

DEPARTMENT OF THE ENVIRONMENT (1986) The Government's First Stage Response to the Environment Committee's Report on Radioactive Waste. London: Department of the Environment.

HOUSE OF COMMONS ENERGY COMMITTEE (1984) Energy Research, Development and Demonstration in the United Kingdom, HC Paper 585, London: HMSO.

HOUSE OF COMMONS ENVIRONMENT COMMITTEE (1986) Radioactive Waste, HC Paper 191-1, London: HMSO.

KASPERSON, R. E. (1984) Ethical Issues in Radioactive Waste Management, Worcester, MA: Centre for Technology, Environment and Development, Clark University.

LEE, T. R. and BROWN, J. M. (1985) 'Research on public attitudes towards nuclear power and radioactive waste management', in House of Commons Environment Committee, Radioactive Waste. HC Paper 253 XVl pp. 536-540.

NECTOUX, F., BALDOCK, D. and BARWISE, J. (1985) Improving Prospects for Jobs in West Cumbria, London: Earth Resources Research.

OPENSHAW, S. (1986) Nuclear Power: Siting and Safety, London: Routledge and Kegan Paul.

O'RIORDAN, T. (1985) 'Radioactive waste disposal: public attitudes, political consequences, in House of Commons Environment Committee, Radioactive Waste. HC Paper 253 XVl, 529-535.

O'RIORDAN, T., KEMP, R., PURDOE, M. P. (1985) 'How the Sizewell B Inquiry is grappling with the concept of acceptable risk'. Journal of Environmental Psychology, 5, (1) 69-85.

van de PLIGT, J. (1985) 'Public attitudes to nuclear power: salience and anxiety'. Journal of Environmental Psychology, 5 (1), 87-98.

NOTES ON CONTRIBUTORS

JENNIFER ARMSTRONG was funded by the Nuffield Foundation to monitor the Sizewell Inquiry for the Town and Country Planning Association. She published her report on this in 1985, and is now a planning consultant. She is on the Executive of the TCPA, and the Royal Town Planning Institute for the East of England.

ANDREW BLOWERS is Professor of Social Sciences at the Open University and has been a Bedfordshire County Councillor since 1973. His publications on environmental politics, planning theory and policy, and local government and politics reflect a combination of academic interest and political experience. His books include The Limits of Power (Pergamon, 1980) and Something in the Air (Harper and Row, 1984), a study of pollution politics. He is currently researching nuclear waste and presented evidence to the House of Commons Select Committee on the Environment on this subject in 1985.

MIRIAM J. BOYLE is Senior Lecturer in Geography at De La Salle College, Middleton. She is interested in cognitive mapping, and in nuclear energy issues and has published a number of papers in these areas.

JOHN BRADBEER is Senior Lecturer in Geography at Portsmouth Polytechnic. His main research and teaching interests are in resource management and environmental planning, in particular energy and minerals, and in recreation. He was a member of the Southern Electricity Consultative Council from 1978 to 1984.

ALAN CRAFT is a Consultant Paediatrician at the Royal Victoria Infirmary, Newcastle upon Tyne. He specialises in the management of children's cancer and is clinically involved in the management of all children who develop cancer throughout the North of England. His academic interests are in the epidemiology of childhood cancer, particularly looking for clues to its causation. He trained in Newcastle and London.

JOHN FERNIE is Senior Lecturer in Economic Geography at Huddersfield Polytechnic. He has taught and researched extensively in North America. His main interests are in energy resource management, especially in nuclear power in the US and UK. He is author of A Geography of Energy in the UK and joint author of Resources, Environment and Policy.

JAMES H. JOHNSON, Jnr. is Associate Professor of Geography at the University of California at Los Angeles. He researches human responses to technological hazards, especially nuclear power plant accidents, and he has testified before the Nuclear Regulatory Commission's Atomic Safety and Licensing Board in emergency planning proceedings on the Diablo Canyon and Shoreham nuclear power plants.

DAVID LOWRY is a member of the Open University's Energy Research Group and a Research Associate with the European Proliferation Information Centre (EPIC) in London. He has written widely on nuclear issues, specialising in the 'back end' of the nuclear cycle; including plutonium, proliferation and waste management. He is co-editor of the six-volume Issues in the Sizewell B Inquiry, published by the Centre for Energy Studies (1982), and submitted evidence in 1985 to the House of Commons Select Committee on the Environment on radioactive waste disposal.

SALLY MACGILL is a lecturer in the School of Geography, University of Leeds. She has teaching and research interests in environmental risk perception, assessment and management, with particular reference to the locational hazards of nuclear and non-nuclear facilities.

STAN OPENSHAW is a lecturer in the Department of Geography, Newcastle University. His research interests include the application of computer modelling and spatial analysis to nuclear war and nuclear power. He is co-author of Doomsday: Britain After a Nuclear Attack, and author of Nuclear Power: Siting and Safety. In 1980 he gave evidence before the House of Commons Select Committee on Energy.

TIMOTHY O'RIORDAN is Professor in the School of Environmental Sciences at the University of

East Anglia. He has written many books and papers on environmentalism, environmental issues and politics and environmental planning and management. He has also been involved practically with environmental matters, for example, by representing the Countryside Commission on the Broads Authority. His best-known books include Environmentalism, Perspectives on Resource Management, The American Environment: Perceptions and Policies (edited with J. Watson) and An Anotated Reader in Environmental Planning and Management (edited with R. Kerry Turner).

M. J. PASQUALETTI is Associate Professor of Geography at Arizona State University. He researches the relationships between land use and nuclear, solar, coal and geothermal energy development. He is co-editor of Nuclear Power: Assessing and Managing Hazardous Technology. He has been Project Director on energy education for the US Department of Education and the National Science Foundation, and is a member of the Arizona Council on Energy Education.

DAVID PEPPER is Principal Lecturer in Geography at Oxford Polytechnic. He researches and teaches environmentalism and war and peace studies. He is the author of Roots of Modern Environmentalism and co-editor of The Geography of Peace and War.

SIÂN PHIPPS is a research student in the School of Geography, University of Leeds, undertaking case study research into the public perception of risk.

W. CLARK PRICHARD is a Waste Management Project Manager at the US Nuclear Regulatory Commission. He works in the area of institutional issues and the regulation of nuclear waste management; and has published papers on waste management and social, demographic and economic impacts of nuclear power plants.

MARVIN RESNIKOFF is a Staff Scientist and Co-Director of the Sierra Club's Radioactive Waste Campaign. Dr. Resnikoff was previously a Project Director at the Council on Economic Priorities, where he wrote The Next Nuclear

Gamble, Transportation and Storage of Nuclear Waste. He is a consultant to the Town of Wayne, New Jersey, in reviewing the clean-up of a local thorium waste dump, to the Southwest Research and Information Center on Shipments of high level waste to the SIPP, and to Spotsylvania County, Virginia, on nuclear fuel transport. He was part of an international team of experts to review, for the State of Lower Saxony, the plans of the nuclear industry to locate a reprocessing and waste disposal operation at Gorleben, West Germany.

MIKE ROBINSON is a Lecturer in Geography at the University of Manchester. He and his wife, MIRIAM BOYLE, share an interest in cognitive mapping and in nuclear energy issues. Since 1981 they have written a number of papers dealing with nuclear energy in France and also with the conduct of the nuclear debate.

JOHN CAMERON STEWART is a Waste Management Project Manager at the US Nuclear Regulatory Commission. He at present works on institutional issues and the regulation of nuclear waste management, and has published papers on waste management and remote sensing. He is a member of the American Institute of Certified Planners.

DONALD J. ZEIGLER is Associate Professor of Geography at Old Dominion University in Norfolk, Virginia, and President of the Virginia Geographical Society. Results of his research on the Three Mile Island and Shoreham nuclear power stations have been widely published and he has testified before the US NRC's Atomic Safety and Licensing Board and at the Sizewell B Inquiry. He is co-author of the AAG monograph, Technological Hazards.

INDEX